Chemical Applications of Far Infrared Spectroscopy

Count that day lost, whose setting sun
Sees no new compound made, or spectrum run

Chemical Applications of Far Infrared Spectroscopy

ARTHUR FINCH, P. N. GATES, K. RADCLIFFE
Royal Holloway College, Surrey, England
F. N. DICKSON
Mellon Institute, Pittsburgh, Pennsylvania, U.S.A.
F. F. BENTLEY
Wright-Patterson Air Force Base Dayton, Ohio, U.S.A.

1970
ACADEMIC PRESS
London and New York

ACADEMIC PRESS INC. (LONDON) LTD

Berkeley Square House
Berkeley Square
London, W1X 6BA

U.S. Edition published by

ACADEMIC PRESS INC.

111 Fifth Avenue
New York, New York 10003

Library of Congress Catalog Card Number: 70–119610
SBN 12–256350–6

PRINTED IN GREAT BRITAIN BY
SPOTTISWOODE, BALLANTYNE AND CO. LTD.
LONDON AND COLCHESTER

PREFACE

Vibrational spectroscopy—infrared and Raman—is far from new; it is, bearing in mind the perspective imposed by the rapidity of scientific advance, an old-established, classic technique for which the relevant theoretical basis was largely laid down very long ago. Hence, new developments necessarily lie in its application to increasingly diverse problems, a use which is dependent on the availability of adequate, and sometimes new, instrumentation. Two outstanding examples of the latter, which have occurred roughly within the last decade, are the utilization of the laser source for Raman work, and the design of spectrometers which operate in the far infrared region.

This book was originally requested in the form of a monograph by a single author, dealing with the applications of far infrared spectroscopy in inorganic chemistry. However, little agreement was found upon a definition of "inorganic", and, therefore, the more encompassing present title was suggested. At the same time, it became apparent that the increased scope demanded more knowledge and experience than that possessed by a single author, and the idea of an edited book was canvassed. This also was rejected, in order to give each author, and not merely one editor, the opportunity to assist in each chapter and hence to reduce to some extent the distractions of vastly different styles and outlooks. Whether this has been successful is for the reader to judge, but certainly each chapter has been read, criticized, (often intemperately) and revised by other than the prime author, such that it would now be misleading to inscribe a single name at the head of the majority of contributions.

Spectroscopists, both in industry and in academic institutions, vary enormously in their requirements, training and outlooks, even when their activities may be described as chemical. Hence there is no apology for the very different approaches used in different portions of this book, the general purpose of which is to encourage scientists to take advantage of the low-frequency portion of the vibrational spectrum. Clearly, it is quite impossible, and not even desirable, to attempt a comprehensive coverage, and although efforts have been made to avoid a simple reflection of the authors' own interests, such influence must be apparent.

A major part of the book was finished during the tenure of a Visiting Fellowship at the Mellon Institute of Carnegie-Mellon University during the summer of 1968, and it is a pleasure to acknowledge the hospitality and expert advice so freely given, and especially by Dr. G. L. Carlson and Dr. F. Kiviat. To Dr. W. G. Fateley, the authors are totally indebted for his constant encouragement, discussion and constructive criticism. A further debt is gladly acknowledged to members of the Analytical

Branch, Materials Physics Division, of Wright-Patterson Air Force Base and to Mr. W. Feairheller of the Monsanto Research Corporation, Dayton, Ohio, and also to Dr. D. Steele and Mr. B. F. Smethurst of Royal Holloway College. Responsibility, of course, rests on the authors, and not on the many friends, including those not mentioned above, who contributed in so many different ways. Partial support from the United States Air Force Materials Laboratory on Contract F33615–68–C–1369 is also gratefully acknowledged.

January 1970 A.F.

CONTENTS

Chapter I

INTRODUCTION

I. Introduction

Although diffraction gratings were first used to obtain infrared spectra as early as 1910, most infrared spectrometers employed prisms as dispersing systems until the present decade; and, hence, the frequency range available was necessarily limited by the optical properties of prism materials. Since most instruments used sodium chloride or potassium bromide optics (useful transmission limits 650 and 400 cm^{-1}, respectively), the region of frequencies lying below about 400 cm^{-1} was initially described as the "far infrared". With the subsequent availability of caesium bromide and caesium iodide prism monochromators, the term was then applied to the region below about 200 cm^{-1}. At the low-frequency end of the scale, a limiting figure of 10 cm^{-1} was usually taken to distinguish from that region of the spectrum normally studied by microwave techniques.

Clearly there are no fundamental (although several important practical) differences between the regions mentioned (10–200 or 400 cm^{-1}) and the adjoining frequency ranges and hence, no precise definition of the term "far infrared" is possible. A recent recommendation of the International Union of Pure and Applied Chemistry suggests the (arbitrary) figures of 10–200 cm^{-1}; that is, the region above the microwave but below the range of prism monochromators. In this book, we shall use the term to cover the 10–400 cm^{-1} region. This is suggested because many absorptions lying between 200 and 400 cm^{-1} are of considerable, especially chemical, interest and, as pointed out in an excellent recent review (Brasch *et al.* 1968) many low-frequency results are difficult to interpret unless the higher range is also considered. Further, wavelengths longer than that represented by 400 cm^{-1} are still less commonly available with conventional commercial spectrometers; the adjective "far" may then fairly be applied.

Far-infrared spectroscopy is not new. Studies in this region started in the early part of the century, as shown by the comprehensive bibliography compiled by Palik (1960), and measurements on the rotational spectra of hydrogen halides (Czerny 1925, 1927), ammonia (Badger 1927, 1928, 1929) and water vapour (Bailey 1930) were made over forty years ago. However, most of these measurements were made with specialized equipment, often operated under conditions of considerable difficulty. Until recently no commercial spectrometers capable of routine operations in the far infrared were available, and the subject attracted relatively few investigators. Within the last decade, major advances in instrumentation have been made in most branches of spectroscopy, and not least in the two divisions (Raman and far infrared) dealing directly with the range 10–400 cm^{-1}. Low-frequency modes have been recorded by Raman techniques for over forty years.

II. Comparison with Raman Spectroscopy

It is appropriate to mention Raman spectroscopy at this stage. This is best regarded as a complementary, rather than an alternative, technique, since the selection rules which govern transitions between vibrational and rotational energy levels are different for Raman scattering and infrared absorption. Even where a lack of symmetry would formally allow the appearance of bands in both infrared and Raman spectra, there is no connexion between intensities in the two spectra, and the detection of a fundamental which is weak or overlaid in, say, the Raman effect may depend upon its location by infrared measurement, and *vice versa*. Ideally, both techniques are used in conjunction where a complete assignment of all vibrational modes is required.

As an example of the comparative uses of far-infrared and Raman spectroscopy, the two boron halides, boron tribromide and tri-iodide, may be considered. In each molecule three of the four fundamental modes of vibrations are expected to lie below 400 cm^{-1}. A satisfactory complete analysis of the tribromide was made in 1936 as a result of Raman measurements alone by Anderson *et al.* (1936); the Raman-inactive mode, ν_2, being located by observation of the Raman-active overtone $2\nu_2$. Boron tri-iodide, however, is chemically unstable and exposure to Raman sources results in considerable decomposition and heavy discoloration due to liberated iodine. At the same time, vibrational analysis using mid-infrared spectroscopy is inadequate, and the use of far-infrared techniques is therefore essential.

Far-infrared spectra of this compound, as a solution in carbon disulphide, and also in benzene, are shown in Fig. 1 (Finch *et al.* 1965). As may be seen, the spectra agree well above 200 cm^{-1} but differ greatly in the region below this point. In carbon disulphide solution there is weak absorption near 189 cm^{-1}, while in the benzene solution this band appears with very considerable intensity, although with little change in frequency. The particularly interesting point is that this band appears at the expected position for the Raman-active (and infrared-inactive)

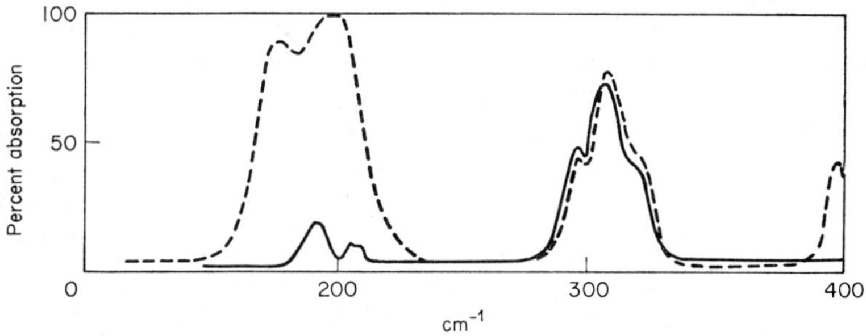

FIG. 1. Far-infrared spectra of boron tri-iodide in (a) benzene solution (-----) and (b) carbon disulphide solution (——). By kind permission of The Faraday Society.

symmetrical stretching vibration ν_1. This may be due to a loss of planarity as the tetrahedral, sp^3, configuration is assumed upon complexing with the solvent; alternatively, some other type of perturbation, leading to relaxation of the selection rules, may be operative. A somewhat similar example is the appearance at about 174 cm^{-1} of the (normally infrared-inactive) I_2 stretching mode in pyridine-iodine solutions (Plyler 1959) and other examples of site-symmetry induced relaxation of the selection rules are known.

In the last few years the potential of Raman spectroscopy has been increased enormously by the introduction of laser sources for excitations, and modes below 20 cm^{-1} have been recorded. It has made possible the successful study of many coloured compounds by suitable choice of exciting frequency. The examination of solid samples, especially powders, has become vastly easier through the use of such sources and some very high resolution studies on gases have been made. An inherent disadvantage is the possible thermal or photolytic decomposition of the sample in the laser beam—an effect which is very much less probable in the infrared—although this may be minimized by the use of specially designed refrigerated cells. On the other hand, Raman methods are eminently suitable for the study of aqueous solutions, which are quite inadmissible in the infrared because of massive absorption in this region of the spectrum.

III. Radiation Sources; Interferometry

The greatest single difficulty in the study of far-infrared spectroscopy by grating monochromators is the lack of a really efficient source of radiation. Compared to the mid infrared, the intensity of radiation from a hot body in the far infrared is small. Hence, in order to isolate the lower from the higher frequencies (to avoid the appearance of spurious bands due to scattered radiation), it is necessary to filter out the higher frequencies. This is often an elaborate process if coverage of the entire range 10–400 cm^{-1} is required and may involve the use of several gratings and various changes and combinations of filter materials. The discovery of an efficient source of radiation in this range, free from higher frequencies, would represent an instrumental breakthrough corresponding to the exploitation of laser sources for Raman spectrometers. At present, the high pressure mercury lamp with quartz windows is the most popular source for measurements below 200 cm^{-1}.

The high frequency limit of the microwave technique was formerly regarded as about 10 cm^{-1} (3.10^5 Mc sec^{-1}) but, using new techniques for harmonic generation, the range has been extended upwards to about 20 cm^{-1} (6.10^5 Mc sec^{-1}) at the present time. Microwave spectroscopy can only be used to study gaseous samples and this limits the general application considerably. Much research is currently in progress in search of new suitable sources for this region.

However, in the absence of dramatic advances in this direction, recent instrumentation research has been successfully concentrated not only on improvements in sensitivity and general performance of detectors, filters, and so on, but also on the application of a different technique, interferometry. Such an application was suggested many years ago by Michelson (1927) but has only become a practical proposition within the last few years. In this technique, the infrared radiation is split into two beams which are eventually recombined and passed through the sample. The path length between the two beams is varied by a moving mirror and the resulting signal from the detector is a function of the energy of the source as modified by any sample absorption. The output is called an interferogram, the Fourier transform of which yields an absorption spectrum; this latter process is carried out on a high-speed digital computer. A large energy advantage over the grating monochromator is that the whole frequency range is simultaneously incident upon the detector rather than a single resolution width. The chief disadvantage in the past has been the time delay caused by the computer processing of the interferogram; however, with the development of relatively small and inexpensive computers designed especially for this task, this is no longer an important consideration. The interferometric method has rapidly gained in popularity as many of the earlier difficulties are disappearing, and is further discussed in Chapter 2.

IV. Units and Energetics

A profusion of different units is used in the low-frequency region to characterize electromagnetic radiation and to measure energy. It is

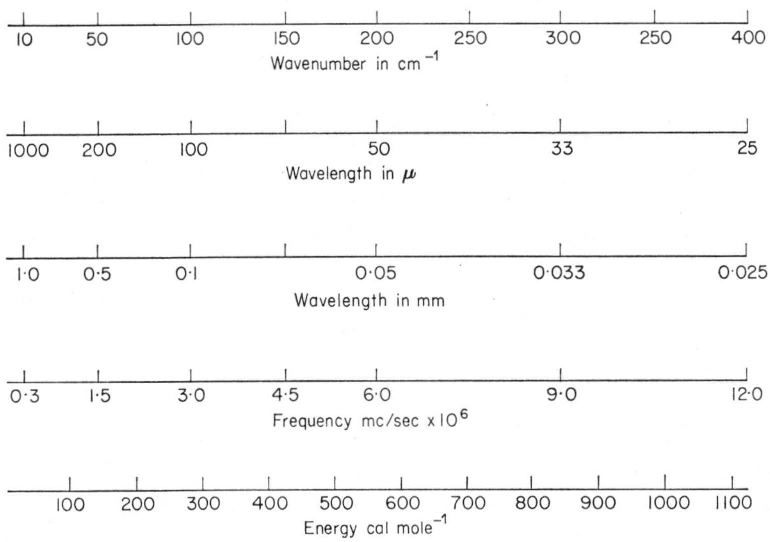

FIG. 2. Correlation chart of units used in the far infrared.

hence useful to tabulate the different units and indicate their relation to one another; this is shown as a correlation chart in Fig. 2.

In this book, the unit most often used will be that of wavenumber, $\bar{\nu}$ (cm^{-1}), that is, reciprocal wavelength, and, following common practice, the term "frequency" will be used in this connexion. This is, of course, not to be confused with absolute frequency, ν, defined by $\nu = c/\lambda$, where c is the velocity of light.

The magnitudes of energy changes in this region are of particular interest. Applying the Planck relation:

$$\Delta E = h\nu = h\bar{\nu}c$$

where ν is the absolute frequency in c $\sec^{-1}$, c is the velocity of light 2.99×10^{10} cm $\sec^{-1}$, $\bar{\nu}$ is the reciprocal wavelength $1/\lambda$ expressed in cm^{-1} and h is Planck's constant, 6.62 erg sec molecule^{-1}; it is possible

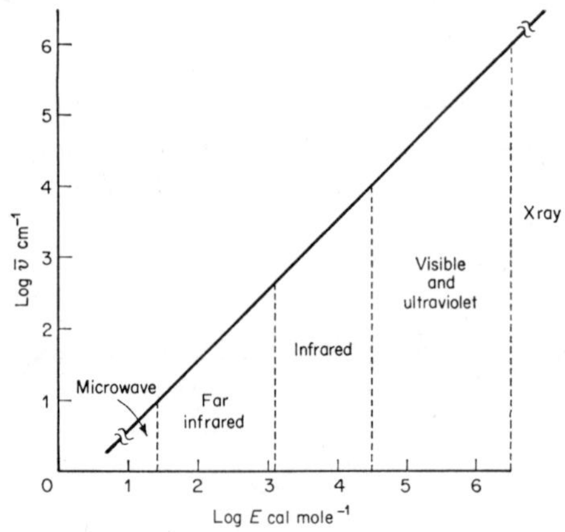

FIG. 3. Energy equivalents associated with various spectral regions.

to calculate energy differences (ΔE) between various vibrational and rotational states. Thus

$$\Delta E \text{ (cal mole}^{-1}) = 2.858\bar{\nu} \text{ (cm}^{-1})$$

and hence the energy equivalent. Even at the upper limit (400 cm^{-1}) of the far-infrared region this is about 1.1 kcal, which is comparable to the translational energy of one mole of an ideal gas ($\frac{3}{2}RT$, or about 1 kcal) and it is less, by a factor of five or so, than the strength of an average hydrogen bond. Clearly then, far-infrared spectroscopy is in principle suited to the study of inter- and intra-molecular displacements at well below the limits of chemical decomposition. A comparison of the magnitudes of the energy equivalent associated with various spectral regions is shown in Fig. 3. At 10 cm^{-1} (that is a wavelength of 1 mm), the energy equivalent is about 0.03 kcal mole^{-1}.

V. Population of Higher Vibrational Levels

In the far-infrared region, transitions starting from vibrational levels other than the ground state are more likely than in regions of higher frequency.

Writing the usual Boltzmann expression we obtain:

$$\frac{n_2}{n_1} = \frac{g_2}{g_1} \exp\left[-\frac{(E_2 - E_1)}{kT}\right]$$

where n_2 and n_1 are the numbers of molecules in two energy levels E_2 (upper) and E_1 (lower), respectively, and g_2 and g_1 are the relevant

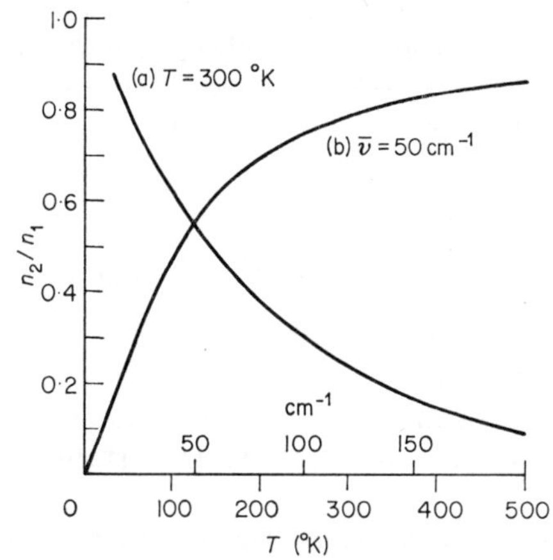

FIG. 4. Relative populations of vibrational levels as a function of (a) wave-number and (b) temperature.

degeneracies of the levels. Hence, putting $g_2/g_1 = 1$ for a non-degenerate vibration and since $E_2 - E_1 = h\nu$, then

$$\frac{n_2}{n_1} = \exp\left(-\frac{hc\bar{\nu}}{kT}\right) = \exp\left(-\frac{1.43\bar{\nu}}{T}\right)$$

where $\bar{\nu}$ is the frequency in cm^{-1} and T is the absolute temperature. Simple substitution for $\bar{\nu}$ at room temperature (300°K) indicates the substantial values for n_2/n_1 below 100 cm^{-1} (see Fig. 4 and Table I).

Thus "hot bands" resulting from transitions from upper levels will be an expected feature of such low-frequency spectra observed at room temperature. Low-frequency vibrations are often distinctly anharmonic: the spacing of the vibrational levels departs markedly from the equal spacing predicted by the simple harmonic oscillator model. Transitions which originate from these upper vibrational levels may thus be considerably shifted from the position of the ground state band.

TABLE I. Relative Population of Vibrational States as a Function of Wave Number.

$\bar{\nu}$ (cm^{-1})	10	100	200	300	400	500	600
n_2/n_1 ($T = 300$°K)	0.95	0.62	0.38	0.24	0.15	0.09	0.008

This effect may obviously be reduced by appropriate reduction of the temperature, which nearly always results in sharpening of the spectra. Depending upon the position in the spectrum, extreme cooling may be necessary, and for a frequency of 50 cm^{-1} even at liquid nitrogen temperatures (77°K) the value of n_2/n_1 amounts to 0.4. Table II lists several calculated values for this effect.

TABLE II. Relative Populations of Vibrational Levels as a Function of Temperature.

T (°K) ($\bar{\nu} = 50$ cm^{-1})	50	77	100	200	400	500
n_2/n_1	0.24	0.40	0.49	0.70	0.84	0.87

VI. Thermodynamic Functions

If a complete assignment of all the fundamental frequencies can be made, and if all molecular structural parameters are known, then the calculation of accurate statistical thermodynamic functions is possible. Low-frequency spectra are particularly important in this connexion, not only because many modes, for example heavy-atom or bending vibrations, occur in this region, but also because the form of the relevant equations is such that the thermodynamic functions are especially sensitive to the low-frequency contributions. This is shown by consideration of the familiar Maxwell-Boltzman expression for the partition function, f:

$$f = \sum_i g_i \exp[-(hc\bar{\nu}_i/kT)]$$

where the summation is carried out over all fundamental frequencies, $\bar{\nu}_i$. From this it is apparent that the terms containing low values of $\bar{\nu}_i$ have the greatest importance. Specifically, vibrational contributions to the total functions are calculated from observed spectral data, using the simple harmonic oscillator approximation for 1 mole of an ideal gas at atmospheric pressure, as follows:

$$C_p^0 = R \sum_i x_i^2 \exp x_i/(\exp x_i - 1)^2$$

$$(H^0 - H_0^0)/T = R \sum_i x_i/(\exp x_i - 1)$$

$$(G^0 - H_0^0)/T = R \sum_i \ln(1 - \exp(-x_i))$$

and

$$S^0 = [(H^0 - H_0^0)/T - (G^0 - H_0^0)/T]$$

where C_p^0, H^0, H_0^0, G_0^0, S_0^0, T and R have their usual significance and $x = \bar{\nu}_i(hc/kT)$ with $\bar{\nu}_i$ in cm^{-1}. A sketch of the variation of these functions with frequency is shown in Fig. 5, calculated from the above, putting $T = 300°\text{K}$, $x/\bar{\nu}_i = 4.8248 \times 10^{-3}$ and using available tables of Einstein functions (Pitzer 1953).

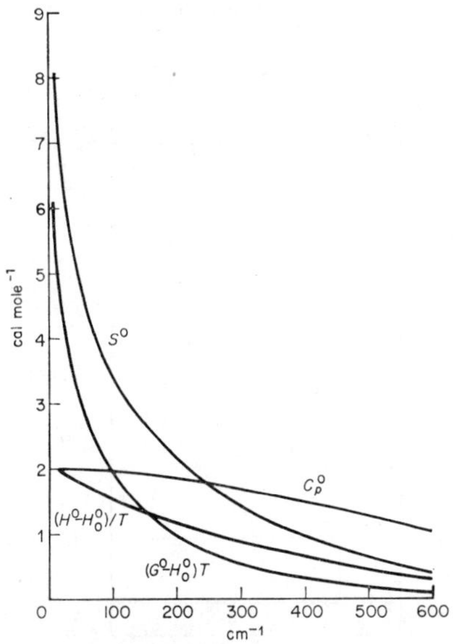

FIG. 5. Plot of statistical thermodynamic quantities as a function of wavenumber.

All functions increase with increasing wavelength, the change being especially sharp with the entropy and the enthalpy function. The total function X_T, where X is any thermodynamic function, is calculated as the sum of all contributions:

$$X_T = x_{\text{translational}} + x_{\text{rotational}} + x_{\text{vibrational}} + x_{\text{electronic}}$$

and since the vibrational contribution, considered above, may typically amount to over 30 per cent of the total figure, the importance of correct assignment of the lower frequencies becomes crucial. Estimates such as the above are based on the usual harmonic oscillator approximation; anharmonicity may be particularly large in low-frequency modes, and appropriate corrections must be applied.

Further aspects of thermodynamic interest are discussed in Chapter 4.

VII. Types of Spectra Observed in the Far Infrared

Various types of far-infrared absorption may be observed and these often depend on the phase state (for example pure rotation in the gas and lattice modes in the crystalline state). The various types can be broadly subdivided into the following classes.

A. Pure Rotational Spectra in the Gaseous State

It is well known that the pure rotational spectra of gases which possess a permanent dipole moment lie in the microwave or far-infrared region.

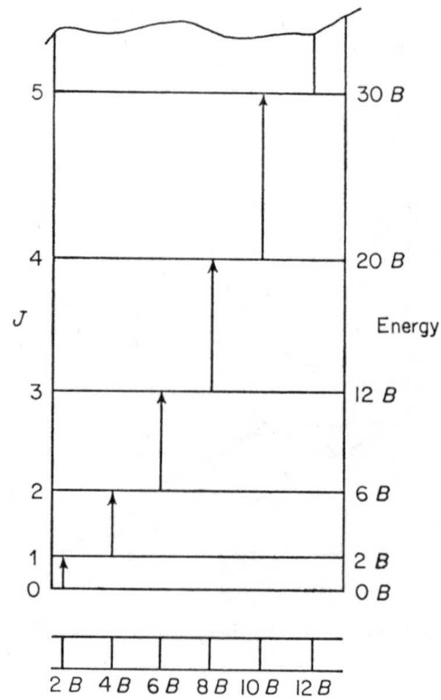

FIG. 6. Energy level diagram showing allowed rotational transitions and spectrum for a rigid diatomic molecule.

Such spectra are usually recorded using microwave spectroscopy, the accuracy and resolution of which considerably exceeds that of far-infrared spectroscopy. Microwave spectroscopy can thus be used to study very detailed features of rotational spectra such as hyperfine splitting and Stark and Zeeman effects, leading to very accurate values of geometrical parameters. This technique is in principle simpler than grating spectroscopy, since instead of a source giving a continuous spectrum over the region required, variable frequency, highly monochromatic Klystron oscillators are used as the source of radiation. Thus, no grating is required and the complications of monochromator systems are avoided; it is essentially an electronic, rather than an optical, technique. At the present time the high-frequency limit of the microwave technique is about 20 cm^{-1} (6.10^5 Mc sec^{-1}).

Taking the simplest example, which is that of a rigid diatomic molecule, the allowed energy levels are given by the expression

$$E_J = \frac{h}{8\pi^2 Ic} J(J+1) \text{ cm}^{-1} \qquad (J = 0, 1, 2, 3 \ldots)$$

where J is the rotational quantum number and I is the moment of inertia about an axis perpendicular to the bond axis.

Putting $(h/8\pi^2 Ic) = B$, the rotational constant, the allowed energy levels are given by:

$$E_J = BJ(J+1) \tag{1}$$

and a simplified energy level diagram representing these is shown in Fig. 6.

The selection rule which governs the transitions among these energy levels in the infrared is $\Delta J = \pm 1$; more specifically, for absorption, $\Delta J = +1$. The allowed transitions are indicated in Fig. 6 and the general expression for the frequencies, ν, of these transitions is easily obtained from eqn (1) by imposing the condition $\Delta J = +1$; that is, by writing $J = (J+1)$ for the upper level and $J = J$ for the lower level. Then

$$\begin{aligned} \nu &= B(J+2)(J+1) - BJ(J+1) \\ &= 2B(J+1) \text{ cm}^{-1} \qquad (J = 0, 1, 2, 3 \ldots) \end{aligned}$$

Hence, a series of lines at frequencies of $2B$, $4B$, $6B$, $8B \ldots$ is observed; and there is a constant spacing of $2B$.

The moment of inertia, I, for such a molecule is given by $I = \mu r^2$ where r is the internuclear distance and μ is the reduced mass of the system, that is $\mu = (m_1 m_2 / m_1 + m_2)$ where m_1 and m_2 are the masses of the two nuclei.

Clearly, the smaller the value of μ, the greater the magnitude of B, and hence the higher the frequencies of the rotational lines. This is illustrated in Table III which compares the frequencies of the first few rotational lines of hydrogen fluoride and carbon monoxide. For many molecules which possess a single non-hydrogen atom (for example HF, HCl, H_2O, NH_3, etc.) all or part of the pure rotational spectrum will lie above the limit of the microwave region, i.e. into the far infrared.

It is found experimentally that with increasing J values, the separation of spectral lines increases from the (constant) value of $2B$ shown in Fig. 6. Since B is proportional to r^{-2}, this represents an increase in the bond length and is interpreted as resulting from the centrifugal distortion

TABLE III. Pure Rotational Lines in Hydrogen Fluoride and Carbon Monoxide.

J	HF[a] $\bar{\nu}$ (cm^{-1})	$C^{12}O^{16}$[b] $\bar{\nu}$ (cm^{-1})
0	41.08	3.845
1	82.19	7.690
2	123.15	11.534
3	164.00	15.379
4	204.62	19.222
5	244.93	23.065
6	285.01	26.907
7	324.65	30.748
8	363.93	34.588
9	402.82	38.426
10	441.13	42.263

[a] Mason and Nielsen (1963). [b] Rao *et al.* (1966).

experienced by the rotating molecule. This centrifugal force, which is proportional to the square of the angular momentum, will be opposed by the restoring force which is calculable in terms of a stretching force constant, k, from observed vibrational spectral data and an assumed simple harmonic (or anharmonic), representation of the vibration dynamics. The simple equation above is then modified to:

$$E_J\,(\text{cm}^{-1}) = BJ(J+1) - DJ^2(J+1)^2 + [HJ^3(J+1)^3 + \ldots]$$

where D, the centrifugal distortion coefficient, $= h^3/32\pi^4 I^2 r^2 kc$. The terms in the square bracket are included to allow for anharmonicity, H etc. being small constant terms dependent on molecular structure. For a simple harmonic oscillator, $D = 4B^3/(\bar{\nu}^2)$ and the weighting of the centrifugal term will increase with decreasing frequency, and the ratio of D values at 10, 100 and 1000 cm^{-1} is $1:10^{-2}:10^{-4}$. Since the first and higher lines may be measured with extreme accuracy using microwave methods, frequencies calculated from such measurements may be used to calibrate far-infrared spectra. Suitable results are listed in Chapter 3.

Water represents a special case; the three principle moments of inertia of the V-shaped molecule are unequal, and result in a very complex rotational spectrum extending over the whole of the far-infrared region as defined in this text. This spectrum invariably intrudes unless special precautions, vacuum operation or meticulous flushing with dry gas, are taken and can serve as a useful secondary calibration source; further details are given in Chapter 3.

B. Intramolecular Vibrations

1. *General*

With the general availability of commercial far-infrared spectrometers and interferometers, there has been extensive application of the techniques to problems of chemical interest, especially in the field of inorganic chemistry. Consider the classical equation for the vibration of a diatomic molecule, using the simple harmonic oscillator approximation:

$$\bar{\nu} = \frac{1}{2\pi c}\sqrt{\frac{f}{\mu}}$$

which relates the fundamental vibration frequency, $\bar{\nu}$, expressed in wavenumbers, to the force constant, f, and the reduced mass, μ, c being the velocity of light. Stretching frequencies in the far-infrared region will then be expected depending upon the value of f/μ. Clearly, this will obtain if either, (a) the value of the reduced mass is large, that is, one or both of the atoms is heavy, or (b) if the force constant is relatively small compared with that found for a normal "chemical" bond. Similar considerations apply to the more complicated normal modes of polyatomic molecules. Both considerations are commonly satisfied in inorganic chemistry, the former especially with heavy-metal to metal, and also metal-to-ligand, bonds, and the latter with the relatively weak bonds found in donor–acceptor complexes of various types. It is obviously not possible to separate completely the two factors of (a) and (b) but in Table IV some examples are listed in which the low-frequency stretching modes derive largely from heavy masses. Stretching force constants in these systems are in the range about 1–3 mdyn Å^{-1}.

TABLE IV. Some Infrared-active Stretching Frequencies in Systems Containing Heavy Atoms.

Compounds	(cm^{-1})	Approximate description of mode
$SnBr_4$	280	antisymmetric Sn–Br stretch
SnI_4	219	antisymmetric Sn–I stretch
$[ZnI_4]^{2-}$	165	antisymmetric Zn–I stretch
$[UCl_6]^{2-}$	260	antisymmetric U–Cl stretch
$ClHgMn(CO)_5$	184	antisymmetric Hg–Mn stretch
$Os_3(CO)_{12}$	121	antisymmetric Os–Os stretch

Low-frequency stretching modes in some hydrogen-bonded systems provide an example of the effect of a small force constant. The H...O stretching frequency in formic acid dimer is at 248 cm^{-1}, the corresponding stretching force constant being at about 0.33 mdyn $Å^{-1}$ (Carlson *et al.* 1966). Similarly, a band at 212 cm^{-1} in ice is associated with H...O stretching, the appropriate force constant being about 0.2 mdyn $Å^{-1}$.

In addition, even if stretching frequencies are available from measurements made in high-frequency regions, other modes (for example, bending modes) may be of considerably lower frequency. An example of this is shown in Fig. 7, where the fundamental frequencies of four phosphorus trihalides are given in correlation-chart form. Although each molecule is of similar (pyramidal) shape and (C_{3v}) symmetry, prediction of the position of fundamentals in one molecule by extrapolation from

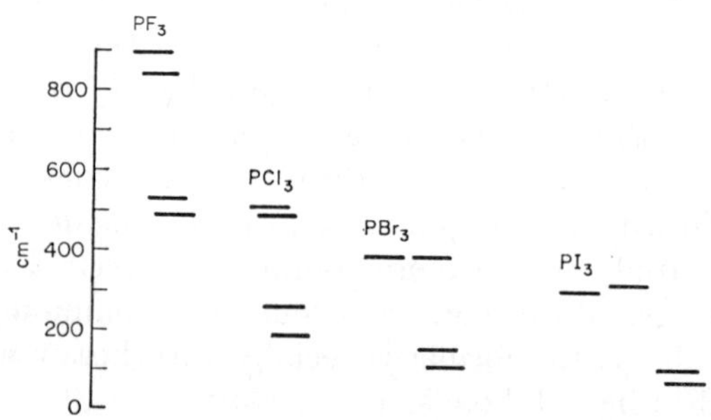

FIG. 7. Correlation chart of fundamental frequencies of the phosphorus trihalides. PF_3 from Cabannes and Rousset 1933; PCl_3, PBr_3 from Davis and Oetjen 1958; and PI_3 from Stammreich *et al.* 1956.

those in another is not possible, and direct measurement is essential. Raman and far-infrared (including interferometric) techniques have both been applied to the study of this system, but complete agreement has not yet been reached (Manley and Williams 1965). As was stressed earlier, calculation of thermodynamic quantities requires values for all fundamental modes, and the low-frequency contributions are especially significant.

2. *Torsional Vibrations*

Torsional modes have also been extensively investigated using far-infrared methods and are discussed in further detail in Chapter 4. In liquids and gases the torsional mode may be the only low-frequency absorption, and, provided the intensity is sufficient, identification is relatively easy. In solids, however, assignment is often complicated by the presence of bands due to lattice vibrations. Torsional vibrations occur because, unless there is free rotation about a given bond, there must exist a potential barrier which hinders the rotation. In favourable circumstances, the height of this rotation barrier may be calculated from the moment of inertia and the observed torsional frequency. Some representative frequencies, ν, and barrier potentials, V, are collected in Table V to serve as examples.

TABLE V. Some Torsional Frequencies ν (cm^{-1}) and Barriers to Rotation V (cal $mole^{-1}$).

	ν	V
CH_3CHO	150	1180
CH_3CDO	111	1166
C_6H_5CHO	111	4660
$(CH_3)_2O$	242	2624
$(CH_3)_2S$	182	2095
o-$Cl(C_6H_4)CH_3$	150	1560

3. *Ring Puckering*

Both far-infrared and microwave studies are relevant to the study of ring conformations and associated vibrational modes, such as out-of-plane bending or puckering, and considerable attention has been devoted to four- and five-membered ring systems. In such near-planar rings the conformation will depend on the relative magnitude of the forces due to ring strain, which will tend in molecules such as, for example, cyclobutane to induce planarity, and on those forces due to hydrogen-atom repulsion, which will tend to favour puckering. Hence, the form of the potential function governing the ring-puckering mode is of importance. Several functions have been proposed, one of the simplest and best known being that due to Bell (1945), which is of quartic degree, $V_x = ax^4$ (cf. the simple harmonic oscillator function, $V_x = -ax^2$). The Bell function depends on the fact that, in puckering, the bond lengths and angles are essentially unchanged. It also requires that this vibration can be treated separately from all other vibrations and that there is free rotation about a single bond. The last condition is not fulfilled, but the added correction (usually quadratic) to the potential function is small and may be much smaller than the quartic term for all except the smallest amplitudes.

In four-membered rings the frequency of the puckering vibration is so low that excited vibrational levels are appreciably populated at ambient temperatures. Probably the simplest molecule to consider is cyclobutane, but the ring-puckering vibration is not active in the infrared, and hence attention has been focussed on the analogous molecules of lower symmetry, trimethylene oxide and sulphide; some results for the former are discussed in Chapter 7.

Figure 8 illustrates the form of the puckering mode and also schematically represents the form of the ring-puckering potential (see, for example, Ueda and Shimanouchi 1967a). In (b) the maximum of the periodic potential and the minimum of the quartic potential overlap and the resultant is of the double-minimum curve found for trimethylene oxide. The barrier at the planar configuration is 15 cm^{-1} (Chan *et al* 1966) and is below the first vibrational level, while for trimethylene sulphide it amounts to 274 cm^{-1} (Borgers and Strauss 1966). In (c) the minima overlap, and the resultant is essentially purely quartic, with no double minimum. This has been found for 2,5-dihydrofuran (Ueda and Shimanouchi 1967a) and for 2,5-dihydrothiophene (Ueda and Shimanouchi 1967b). Other forms of potential functions may be employed, for example four-parameter equations which include an exponential term (Ueda and Shimanouch 1967b).

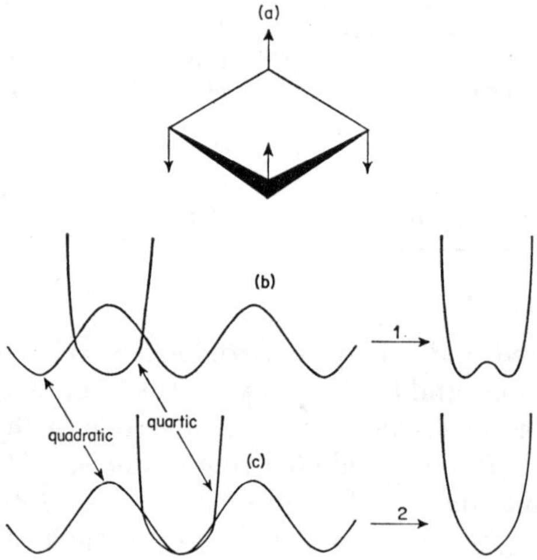

FIG. 8. (a) Ring puckering mode of a four-membered ring; (b) diagrammatic representation of the origin of a double-minimum; and (c) a single-minimum ring puckering potential.

C. Intermolecular Vibrations

This is a rather loose classification covering a range of different modes, including some of the effects specific to one phase; for example, lattice modes in molecular crystals, where one molecule essentially vibrates against another in the unit cell. Further, the distinction between *intra*- and *inter*-molecular vibrations is not always clear. The main types of intermolecular modes may conveniently be subdivided as follows.

1. *Complexes*

Stretching of the link between the two interacting molecules in a weak complex can sometimes be observed in the far infrared. In principle, where two nonlinear molecules are linked together in a complex, a maximum of six new vibrational modes should arise as the separate molecules will have $3N_1$-6 and $3N_2$-6 normal modes (where N_1 and N_2 are the number of atoms in the constituent molecules) and the complex will have $3(N_1+N_2)$-6 normal modes. The number of observable modes may, of course, be reduced by symmetry. An example of the stretching of such a link is afforded by the N–I link in the pyridine/iodine bromide complex at 144 cm^{-1} (Ginn and Wood 1966). Many deductions about coordinate bond properties have been made by considering changes in the vibrational spectra of one of the constituent molecules in a series of complexes. Observations on the coordinate bond stretching mode in the far infrared in a series of complexes should give more direct information about the nature of the bond.

2. *Hydrogen Bonding*

Infrared studies have been described as "*the* most sensitive, *the* most characteristic and *one of the* most informative manifestations of the hydrogen bond" (Hiebert *et al.* quoted by Pimentel and McClellan, 1960). This statement referred to the perturbation in the hydrogen-containing molecule R–A–H of the A–H stretching (ν_s) and the R–A–H in-plane bending (ν_b) modes, both of which are found at high frequencies (2500–3500 cm^{-1} and 1000–1700 cm^{-1} respectively). Much earlier work, including the above, was related to intramolecular hydrogen bonding; where intermolecular bonding is considered, then the system may be treated as a special form of complex formation. Thus, as indicated in the preceding section, formation of the complex will result in the loss of three translational and three rotational degrees of freedom. Of the six new vibrational modes, one, ν_σ, will be directly associated with the stretching vibration of the hydrogen bond itself, R–A–H...B. Values of ν_σ lie in the range 50–250 cm^{-1}, and clearly give direct information about the strength of the bond in terms of the force constant. Thus, an assignment by Miyazawa and Pitzer (1959), based mainly on normal coordinate calculations adjusted to give a best fit to entropy data, put ν_σ for *gaseous* formic acid at 248 cm^{-1}; this was recently confirmed by Carlson *et al.* (1966) using far-infrared measurements. This figure leads to a value 0.33 10^{-5} mdyn Å^{-1} for the stretching force constant, interestingly confirming a prediction of 0.4 10^{-5} dyne cm^{-1} made over twenty years ago by Halford (1946). Similarly, an absorption at 212 cm^{-1} in ice has been correspondingly identified as ν_σ and, using as a very crude model for this system that of a harmonic oscillator with point masses of eighteen, leads to a force constant of a similar magnitude, *viz* 0.2 10^{-5} mdyn $Å^{-1}$ (Pimentel and McClellan 1960).

The other five "new" vibrational modes are deformations, and again are expected to be of low frequency. If the complex exists as a dimer, as with various carboxylic acids, then two degrees of freedom will be associated with the hydrogen-bond stretching, and four with ring deformation modes.

Such studies involve considerable experimental difficulty. Bands are often broad and weak, the phase state and temperature are important variables, and complementary Raman measurements are often essential. Very roughly, the frequency positions relate to the potential function describing the equilibrium configuration of the complex, and the intensity behaviour reflects the movement of charge associated with perturbation of the equilibrium state. Since the hydrogen bond is of such importance in biologically interesting molecules, these studies are of current interest. In particular, the low-frequency modes contribute greatly to the entropy of the hydrogen bond and are necessary data in calculations of equilibrium constants. Further discussion of this topic is given in Chapter 5.

3. *Lattice Modes*

(a) *General*

Crystal spectra exhibit two types of vibrations: internal and external or lattice modes. Internal modes are usually found at frequencies very close to those observed in the liquid or gas phase, but the intensities are often markedly different, and such spectra are rarely preferred for studies directed at establishing the structure of molecular units. Lattice modes involve the relative motions of molecules or complex ions as a whole within the crystal lattice, and may either be of translatory or libratory origin. The forces between molecules are usually an order of magnitude weaker than the inter-atomic forces; and since a molecular grouping is involved, the mass will usually be relatively high. Hence, modes associated with such displacements will be expected in the far-infrared region.

If a unit cell, that is the smallest unit from which the entire lattice may be generated by translation, contains P molecular species, each with Q atoms, there will be a total of $3(PQ\text{-}1)$ crystal modes. Of these, $3Q\text{-}6$, each repeated P-fold, will be intramolecular, and of the remaining

3(2*P*-1), there will be 3(*P*-1) lattice translational modes and 3*P* lattice rocking or librational modes, which will be in the low-frequency region. In addition, there may be absorptions resulting from coupling of internal modes with lattice vibrations of the proper symmetry. Such effects are normally small, but may be significant with low-frequency fundamentals, that is bending modes involving heavy atoms.

The experimental distinction between observed translational and librational modes is of interest. The latter are expected to show greater intensity than the former, especially for molecules with high dipole moments such as the hydrogen halides. Since translational vibrations depend upon the total molecular mass, while librational modes are functions of the moments of inertia, isotopic substitution, especially deuteration, has been used to distinguish these. Thus, in a study of some hydrogen and deuterium halides at liquid nitrogen temperatures (Anderson *et al.* 1963), the following assignments shown in Table VI were made.

TABLE VI. Lattice Modes in Hydrogen and Deuterium Halides (cm^{-1}).

	Translational		Librational		
HCl	86	109	217	296 (doublet)	
DCl	89	113	169	209	258
HBr	57	71	200	269 (doublet)	
DBr	60	72	154	194	233

The two lowest bands in each spectrum were assigned as translational modes; they were sharper and less intense than those at higher frequencies, and the change upon deuteration was small. Interestingly, it may be noted that the isotope shift is to higher frequencies, contrary to expectation, and in view of the unknown crystal structures of these compounds, this is possibly due to different crystal packing. The translational assignment is strengthened by the fact that the ratios of corresponding frequencies in the chloride and bromide respectively are close to the square root of their inverse mass ratio.

The higher frequencies were assigned to librational modes on the basis of greater intensities and isotopic shifts. The maximum value of the ratio of corresponding halide (undeuterated and deuterated) frequencies is $\sqrt{2}$ for librational frequencies, where only the hydrogen atoms move. For this reason the assignment of the weak absorptions at 496 cm^{-1} (HCl) and 400 cm^{-1} (HBr) to librational lattice modes was untenable and it was concluded that the bands at 296 cm^{-1} (HCl) and 269 cm^{-1} (HBr) were doublets containing two fundamentals.

One of the aims of this application of far-infrared measurement is the calculation of intermolecular forces. The values of the translational and torsional force constants, F_a and F_b, for the hydrogen and deuterium halides are shown in Table VII.

TABLE VII.

	HCl	HBr	DCl	DBr
F_a	2.1	2.0	2.2	2.1 dyn cm^{-1} 10^4
F_b	0.42	0.35	0.5	0.43 dyn cm^{-1} 10^4

Values of F_b are roughly an order of magnitude less than F_a, which themselves are correspondingly less than the intramolecular stretching force constants (HCl: 5.2 10^5, HBr: 4.1 10^5 dyn cm^{-1}). The variation of the intramolecular force constants in hydrogen chloride, bromide and iodide decrease from gas to solid by 10.4, 10.4 and 9.8% respectively (Hornig and Osberg 1955). This is surprisingly constant, and much greater than expected from environmental polarizability considerations; presumably it is associated with specific interactions since HBr in a solid nitrogen matrix shows a decrease of only 0.05% (Becker and Pimentel 1956).

(b) *Solid Solutions*

The selection rules for all $3(PQ\text{-}1)$ modes may be derived from the unit-cell symmetry and the selection rules for the intramolecular $3Q$-6 modes can also be obtained from the site symmetry of the molecular species. It can be shown that the site group method is inadequate for polarization values, or for overtone and combination selection rules.

The effects caused by the fact that the potential field at a site in the crystal differs from that acting on a molecule in the gas phase are termed "static field effects", and may produce shifts in frequency, splitting of degenerate fundamentals and the appearance of "forbidden" fundamentals. The coupling of molecules within a unit-cell may further split each fundamental into a maximum of Q components—"correlation field effect".

Use of dilute isotopic solid solutions for distinguishing static field and correlation field effects has been made by Hiebert and Hornig (1952) and Hrostowski and Pimentel (1951). Similarly, the solvent may be a closely related compound, for example, naphthalene-anthracene. The static field for a molecule surrounded by isotopic molecules is considered to be the same as that of a molecule in a pure component. The isotopically unique molecule is then largely uncoupled from its surroundings, and the frequency for a completely uncoupled molecule is expected to coincide with the weighted average of the frequencies of all correlation field components, whether active or not. An excellent review of the use of this technique and the effect of matrix environments upon the frequencies and intensities of vibrational modes is given by Vedder and Hornig (1961).

4. *Polar Interactions*

Within the last five years various groups have reported observations of ill-defined, broad absorption bands in the far-infrared spectra of polar molecules. Jakobsen and Brasch (1964) noted in the spectra of thirteen organic compounds, liquid and solid, such absorptions which, in the range 88 to less than 50 cm^{-1}, could not be assigned to intramolecular, lattice or torsional modes. For some compounds, for example acetonitrile, the absorption persisted into the gas phase, although at lower frequency (about 87–65 cm^{-1}). Since these bands appeared for compounds with appreciable polarity, such as *o*-dinitrobenzene which has a dipole moment of 6 Debyes, but not with similar compounds which were non-polar, such as *p*-dinitrobenzene which has zero dipole moment, it was suggested that their origin lay in intermolecular association, probably due to dipole–dipole interaction. Gebbie and Chantry (1965) found an apparently similar type of absorption in liquid chlorobenzene, which was measured interferometrically in the 10–400 cm^{-1} region. It was shown that sharp absorption, presumably lattice modes, occurred in the same spectral region when the chlorobenzene was examined in the solid state at low temperatures. They therefore assumed that the broad, liquid-phase mode arose from "pseudo-lattice" vibrations.

Recently, Gabelnick and Strauss (1967) have described absorption at less than 50 cm^{-1} in liquid carbon tetrachloride and established that no absorption in that region occurs using gas samples. Carbon tetrachloride is normally regarded as being completely non-polar, with no dipole moment. Since their spectrum resembled that of a pure rotation spectrum, they treated their data, using methods developed by Gordon (1965), as though the CCl_4 molecule had a fixed dipole moment. The results were interpreted as indicating an "effective dipole moment" of 0.1–0.2 D in carbon tetrachloride, which corresponds to bending a C–Cl bond through about ten degrees. However, although the spectrum resembled that of a pure rotation spectrum, the low value of the calculated moment of inertia again indicated intermolecular interactions.

Previous evidence for the existence of intermolecular interaction in polar organic molecules has generally been indirect, as from interpretation of virial data, and this method represents a possible direct approach to the problem

VIII. Reviews

The first outline and discussion of applications of far-infrared spectroscopy was published by Wilkinson *et al.* (1962) and was followed by a review by Wood (1963). Although these were written before the widespread use of interferometers became common, and when few commercial spectrophotometers were available, both are still highly relevant. They have been supplemented by an excellent and very recent review by Miller (1968) which, together with some presently unpublished work, includes an authoritative account of current knowledge on torsional vibrations and barriers to rotation. The approach is to problems of mainly chemical interest; inorganic complexes are not included. The latter, and metal-ligand vibrations, in general, are the subject of publications by Adams (1967) and Nakamoto (1963) which are the best source of detailed information, and there are also specialized monographs by Clark (1965, 1967) which contain valuable data.

A recent article by Ferraro (1968) deals with developments and trends in inorganic far-infrared spectroscopy, and contains information derived from the author's experience in the application of high-pressure techniques. A very comprehensive collection of infrared spectra of both inorganic and organic molecules has been published by Bentley *et al.* (1968). Correlation tables, discussion of characteristic frequencies, and applications to the analysis of a wide variety of materials are presented. The range 700–300 cm^{-1}, approximately, is covered, and may subsequently be extended to lower frequencies. This supplements an earlier article by Stewart (1965) which also deals with the possibilities of establishing functional group characteristic frequency correlations. An excellent literature coverage of work done in chemical aspects of far-infrared spectroscopy is available by Brasch *et al.* (1968) for the years 1965 and 1966; this review also contains some matter published in 1967. An earlier review by Palik (1960) covers the literature up to 1960.

References

Adams, D. M. (1967). "Metal-Ligand and Related Vibrations," 2nd publication, St. Martin's Press, London, references as cited therein.

Anderson, T. F., Lassettre, E. N., and Yost, D. M. (1936). *J. Chem. Phys.* **4**, 703.

Anderson, A., Gebbie, H. A., and Walmsley, S. H. (1963). *Molec. Phys.* **7**, 401.

Badger, R. M. (1927). *Proc. natn. Acad. Sci. U.S.A.* **13**, 408.
Badger, R. M. (1928). *Nature, Lond.* **121**, 942.
Badger, R. M. and Cartwright, C. W. (1929). *Phys. Rev.* **33**, 692.
Bailey, C. R. (1930). *Trans. Faraday Soc.* **26**, 203.
Becker, E. D. and Pimentel, G. C. (1956). *J. Chem. Phys.* **25**, 224.
Bell, R. P. (1945). *Proc. R. Soc.* (A) **183**, 328.
Bentley, F. F., Smithson, L. D., and Rozek, A. L. (1968). "Infrared Spectra and Characteristic Frequencies." Interscience, New York.
Borgers, T. and Strauss, H. L. (1966). *J. Chem. Phys.* **45**, 947.
Brasch, J. W., Mikawa, Y., and Jakobsen, R. J. (1968). "Applied Spectroscopy Reviews." (E. G. Brame, Jr., ed.). Vol. 1. E. Arnold and M. Dekker, New York.
Cabannes, J. and Rousset, A. (1933). *Ann Phys.* **19**, 229.
Carlson, G. L., Witkowski, R. E. and Fateley, W. G. (1966). *Spectrochim. Acta.* **22**, 1117.
Chan, S. I., Borgers, T., Russell, J. W, Strauss, H. L., and Gwinn, W. D. (1966). *J. Chem. Phys.* **44**, 1103.
Clark, R. J. H. (1965). *Rec. chem. Prog.* **26**, 269.
Clark, R. J. H. (1967). "Halogen Chemistry." (V. Gutmann, ed.) Vol. 3. Academic Press, London.
Czerny, M. (1925). *Z. Phys.* **34**, 227.
Czerny, M. (1927). *Z. Phys.* **44**, 235.
Davis, P. W. and Oetjen, R. A. (1958). *J. molec. Spectrosc.* **2**, 253.
Ferraro, J. R. (1968). *Analyt. Chem.* **40**, 24A.
Finch, A., Hyams, I. J., and Steele, D. (1965). *Trans. Faraday Soc.* **61**, 203.
Gabelnick, H. and Strauss, H. (1967). *J. Chem. Phys.* **46**, 396.
Gebbie, H. A. and Chantry, G. W. (1965). *Nature, Lond.* **208**, 378.
Ginn, S. G. W. and Wood, J. L. (1966). *Trans. Faraday Soc.* **62**, 777.
Gordon, R. G. (1965). *J. Chem. Phys.* **43**, 1307, and papers cited therein.
Halford, J. O. (1946). *J. Chem. Phys.* **14**, 395.
Hiebert, G. L. and Hornig, D. F. (1952). *J. Chem. Phys.* **20**, 918.
Hornig, D. F. and Osberg, W. E. (1955). *J. Chem. Phys.* **23**, 662.
Hrostowski, H. J. and Pimentel, G. C. (1951). *J. Chem. Phys.* **19**, 661.
Jakobsen, R. J. and Brasch, J. W. (1964). *J. Am. chem. Soc.* **86**, 3571.
Manley, T. R. and Williams, D. (1965). *Spectrochim. Acta.* **21**, 1773.
Mason, A. A. and Nielsen, A. H. (1963). Scientific Report No. 5, AF 19(604)-7981.
Michelson, A. A. (1927). "Studies in Optics," p. 45. University of Chicago Press, Chicago.
Miller, F. A. (1968). "Proceedings 4th Conference on Molecular Spectroscopy". Hydrocarbon Research Group, Inst. Petroleum (London). To be published.
Miyazawa, T. and Pitzer, K. S. (1959). *J. Am. chem. Soc.* **81**, 74.
Nakamoto, K. (1963). "Infrared Spectra of Inorganic and Coordination Compounds." John Wiley, New York.
Palik, E. D. (1960). *J. opt. Soc. Am.* **50**, 1329.
Pimentel, G. C. and McClellan, A. L. (1960). "The Hydrogen Bond", Chap. 3. Freeman, San Francisco.
Pitzer, K. S. (1953). "Quantum Chemistry." Prentice Hall, Englewood Cliffs, New Jersey.
Plyler, E. K. and Mulliken, R. S. (1959). *J. Am. chem. Soc.* **81**, 823.
Rao, K. N., Humphreys, C. J., and Rank, D. H. (1966). "Wavelength Standards in the Infrared." Academic Press, New York.

Stammreich, H., Forneris, R., and Tavares, Y. (1956). *J. Chem. Phys.* **25**, 580.
Stewart, J. (1965). "Interpretive Spectroscopy." (S. K. Freeman, ed.) Reinhold, New York.
Ueda, T. and Shimanouchi, T. (1967a). *J. Chem. Phys.* **47**, 4042.
Ueda, T. and Shimanouchi, T. (1967b). *J. Chem. Phys.* **47**, 5018.
Vedder, W. and Hornig, D. F. (1961). "Advances in Spectroscopy", Vol. II, p. 217. Interscience, New York.
Wilkinson, G. R., Inglis, S. and Smart, C. (1962). "Spectroscopy", Rept. Conf. organized by Inst. Petrol. Hydrocarbon Res. Group, London. p. 157.
Wood, J. L. (1963). *Q. Rev. chem. Soc.* **17**, 111, 362.

Chapter 2

INSTRUMENTAL CONSIDERATIONS IN THE DESIGN OF FAR-INFRARED SPECTROMETERS

I. Introduction

Essentially, the problems which must be solved in the design and construction of a far-infrared spectrometer are, (a) to provide a source having sufficient energy in the frequency region of interest, (b) to eliminate higher order radiation which might be present with sufficient intensity to be detected, (c) to design a monochromator in which the radiation is sufficiently dispersed to provide adequate spectral resolution, (d) to provide a detector which responds to radiation in the region of interest, and (e) to equip the detector with an appropriate amplifier and recorder.

II. Sources

The greatest single difficulty in the study of far-infrared spectroscopy is the lack of a really efficient source of radiation. Indeed, the practice of far-infrared spectroscopy would be revolutionized by the availability of a tunable coherent source. In recent years coherent output, using lasers, has been achieved at most far-infrared frequencies but little progress has been made on the problem of tuning the laser output. The production of far-infrared radiation by generating harmonics from a microwave klystron has evidently not been a practical proposition. The use of thermal sources, despite the drawbacks mentioned below, is inevitable at present. These sources are assumed to approximate to "black bodies", thereby allowing application of the well-known laws pertaining to such systems.

The first of these, Stefan's law, relates the total power per unit area radiated from a black body, $E(T)$, to its temperature, T (°K) as follows:

$$E(T) = \sigma T^4 \tag{1}$$

where σ is Stefan's constant (5.6687×10^{-12} watt cm^{-2} °K^{-4}).

Secondly, it is necessary to know how the radiant energy is distributed with respect to frequency, ν. This is given by Planck's formula as follows:

$$E(\nu, T) = \frac{2\pi h\nu^3}{c^2(\exp(h\nu/kT) - 1)} \tag{2}$$

where $E(\nu, T)$ is the energy radiated per unit volume per unit frequency interval, c is the velocity of light, h is Planck's constant, and k is Boltzmann's constant. The form of this equation implies a pronounced energy maximum at any fixed temperature, and that the output falls off sharply with decreasing frequency. The position of such maxima may be calculated by differentiating eqn (2), setting $dE(\nu, T)/d\nu = 0$ and solving the resulting equation for ν. This leads to

$$\nu_{\max}\,(\mathrm{cm}^{-1}) = 3.45T\,(°\mathrm{K}) \tag{3}$$

Thus, in the frequency range of 10–400 cm^{-1}, the optimum temperatures would lie between 2.9 and 116°K.

Substitution of eqn (3) into the Planck formula (2) gives an expression for the maximum energy radiated, $E(\nu, T)_{\max}$. This is of the form

$$E(\nu, T)_{\max} = BT^3 \tag{4}$$

where B is a constant. Thus the maximum intensity of the energy emitted is proportional to T^3 and, for the practical purpose of a radiation source, the intensity at the very low temperatures mentioned above would be far too low. The only solution is to increase the temperature considerably, as in a conventional infrared source. This means that the flattened portion of the curve (eqn 2), away from the peak intensity, is being considered, and for low frequencies and high temperature the assumption that $h\nu \ll kT$ is made. Planck's formula then reduces to

$$E(\nu, T) = \frac{2\pi kT\nu^2}{c^2} \tag{5}$$

which is the Rayleigh-Jeans' law and represents adequately the flattened portion of the curve. Thus the radiant energy emitted is now proportional to the absolute temperature and the square of the frequency. However, from Stefan's law, the intensity of the radiation from *all* frequencies is proportional to T^4 and, although by raising the temperature the intensity of the far infrared radiation is increased, the intensity in the unwanted higher-frequency region increases much more rapidly.

From this discussion it is evident that an ideal black body would not be the most suitable type of source of far-infrared radiation. Although early workers used a Globar because of its emission above 200 cm^{-1}, the high pressure mercury arc is now almost universally used due to its superior performance below 50 cm^{-1} (Plyler *et al.* 1962). At frequencies above 100 cm^{-1} the radiation is almost exclusively from the heated envelope acting as a *grey* body, whereas below 50 cm^{-1} "bremstrahlung" radiation from the arc predominates (Genzel and Eckhardt 1954). Papoular (1964) found that above 40 cm^{-1} the apparent temperature of a high pressure mercury arc was constant at about 1500°K while below 40 cm^{-1} it rose steadily to 3000°K at 10 cm^{-1}. In a similar experiment, Cano and Mattioli (1967) demonstrated that the lamp emission approximated to that of an optically thin plasma in that it was independent of frequency below 25 cm^{-1}. Since the energy of a black body is proportional to the square of the frequency (the Rayleigh-Jeans approximation) this low-frequency plasma emission explains the rise in apparent temperature of the arc below 50 cm^{-1}, and its superiority to the Globar.

III. Dispersion

For a spectrometer in which the resolution is limited by source energy or detector sensitivity, as is the case in the far infrared, the energy throughput, E, is given by the equation due to Strong (1949)

$$E = B_0 T\theta\left(\frac{sA}{F}\right)\Delta\nu^2\frac{d\theta}{d\lambda} \tag{6}$$

where B_0 is the emissivity constant of the source, whose effective temperature is θ; A is the cross-sectional area of the dispersive element, which has a dispersion $d\theta/d\lambda$; F is the focal length of the collimated beam; s is the slit length; $\Delta\nu$ is the frequency band passed by the exit slit and T is the transmission efficiency of the instrument. B_0 and θ are dependent on the properties of the source and in order to maximize the energy throughput at a given spectral resolution, the monochromator must be designed to maximize the remaining variables in eqn (6).

It can be shown that gratings are markedly superior to prisms as dispersive elements: in general, gratings can be fashioned with larger areas than prisms and, for a given resolving power, a grating instrument is capable of passing an order of magnitude more energy than a prism instrument. Grating spectrometers do, however, suffer from the disadvantage of producing overlapping higher-order spectra in addition to the one required. Below 200 cm^{-1}, gratings are exclusively used as the dispersive elements in far-infrared spectrometers. The cross-sectional area, A, of the grating is limited by the expense and availability. A further practical consideration is that very large gratings would imply an unreasonably large instrument.

The mirrors employed in far-infrared spectrometers are front-surface aluminized and normally have a very high transmission efficiency, T. It is, therefore, not worthwhile attempting to increase T in order to maximize E.

The remaining two variables in eqn (6) are usually dealt with together in the form of the ratio (s/F). This ratio is at the discretion of the designer, who may either increase s or decrease F. The latter approach requires extremely curved mirrors and critical positioning of both the optical components and the sample, due to the "fast" beam. Increasing s by the use of very long slits normally introduces aberration and vignetting problems. However, these can be negotiated by using the Ebert-Fastie or Czerny-Turner monochromator layouts (Fastie 1952; Czerny and Turner 1930). These two layouts are essentially the same: the single monochromator mirror of the Czerny-Turner system being replaced, in the Ebert-Fastie system, by two mirrors with the same centre of curvature.

The use of the above layout in most high resolution far-infrared spectrometers has resulted from the following factors: (a) aberrations are minimized by the use of curved slits, which may thus be long and pass more energy, (b) the mirror is spherical and hence cheaper and more readily available than, say, parabolic mirrors used in other layouts, and (c) the system is compact and hence facilitates evacuation. Evacuation or very efficient purging of spectrometers operating in the far-infrared spectral region is essential for complete removal of water vapour absorption.

IV. Image-reducer

The maximum energy throughput for a given resolving power is thus realized when the ratio slit height/focal length is maximized by the use of Ebert-Fastie-type optics. In theory, the slit length may be very large (since the aberrations are minimized) but in practice, since detectors employed in the far infrared have small area windows, problems may arise in condensing all the radiation issuing from the exit slit on to the detector window. An elegant method of reducing this problem is the use of an image-reducer (Benesch and Strong 1951) which, in addition to reforming the elongated slit image into a more circular area, also results in an increase in signal/noise ratio for a given resolving power; this will now be demonstrated.

When an exit slit image, of length s, is condensed directly on to the detector window, of height l, then $s = lf_m/f_d$ where $f_m(f_d)$ is the aperture of the monochromator (detector). In the Ebert-Fastie type of optics the monochromator aperture, f_m, is given by $f_m = F/A^{1/2}$, where F is the focal length of the Ebert mirror and A the area of the grating. Hence

$$s = \frac{lF}{f_d A^{1/2}} \tag{7}$$

and substitution of this into eqn (6) yields

$$E = \frac{klA^{1/2}\Delta\nu^2}{f_d} \tag{8}$$

where k is a constant depending upon the properties of the grating and source. This shows that, for a given resolving power, the detector window height, l, limits the energy throughput.

As the detector window is usually circular, while the slit image is elongated, the principle of the image reducer is effectively to increase l by slicing the slit image to produce a series of m parallel images. Equations (7) and (8) now take the forms

$$s = \frac{mlF}{f_d A^{1/2}} \quad \text{or} \quad F = \frac{sf_d A^{1/2}}{ml} \tag{9}$$

and

$$E = \frac{mklA^{1/2}\Delta\nu^2}{f_d}. \tag{10}$$

It can be seen, by substituting eqn (9) into eqn (6), that the energy throughput is increased by a factor of m by reducing the focal length of the monochromator to $(1/m)$ of its original value.

Referring to Fig. 1, which illustrates the principle of the image reducer, it can be seen that in order to produce a square image the maximum slit width usable is given by

$$\omega_{\text{max}} = \frac{s}{m^2} \tag{11}$$

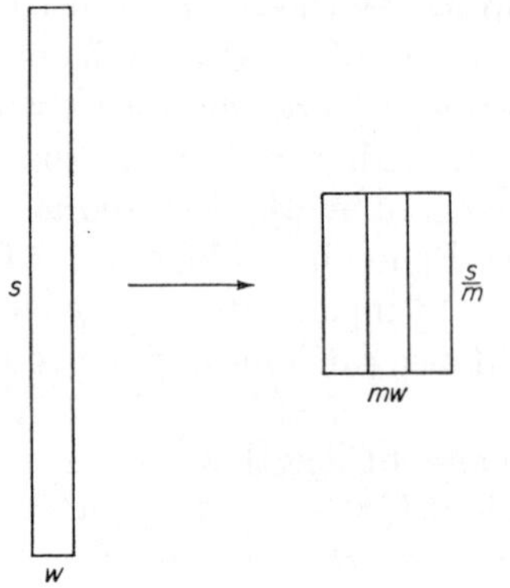

FIG. 1. Principle of image reducer.

Strong (1949) related the slit width in wavenumbers, $\Delta\nu$, to the mechanical slit width in centimetres, ω, as

$$\Delta\nu = \frac{\omega}{F\lambda^2 \, d\theta/d\lambda} \tag{12}$$

Substitutions of eqns (9), (11) and (12) into eqn (6) yields

$$E_{\max} = \left(\frac{B_0 \, T\theta l^3}{f_d \, A^{1/2}\lambda^4 \, d\theta/d\lambda}\right)\left(\frac{1}{m}\right) \tag{13}$$

which shows that for a given assembly of source–detector–grating, the energy throughput at maximum slit is inversely proportional to the number of slices taken.

The choice in the number of slices to take is thus a compromise between the increase in energy obtained at a given resolving power due to an increase in m (eqn 10) and the decrease in the maximum energy available (eqn 13).

V. Filters

A. Filtering

The grating is not a complete monochromator by itself since application of the well-known grating formula in the rearranged form $n = 2\nu d \sin\theta$, where n is the spectral order, d the grating spacing and ν the frequency, shows that for an angle of rotation, θ, of the grating such that the diffracted image falling on the exit slit has a frequency ν in the first order, then in general the nth order at $n\nu$ will also fall on the exit slit. Hence some sort of filtering method must be employed in order to produce pure radiation at the first-order frequency.

The magnitude of this effect was illustrated by Oetjen and co-workers (1952) who considered an ideal echelette grating (180 lines per inch and blaze angle of 21°) which concentrated radiation of 100 cm^{-1} into the first order, when used with a black body source operating at 1300°K. The spectral interval passing the exit slit will be proportional to $(1/n)$ since dispersion is directly related to the order number. Thus the radiant power reaching a detector in a given order will be proportional to the ratio of radiant power of source at the order frequency to order number. Oetjen chose the first-order frequency to be unity and then calculated the relative intensities in the various orders. These are shown in Table I.

TABLE I. The Relative Intensities of Energy Diffracted by an Ideal Echelette Grating into the Various Orders.

Spectral region	Spectrum order	Relative intensities
100 (cm^{-1})	1	1
200	2	8
300	3	24
400	4	54
500	5	99
600	6	161
700	7	241
800–1,000	8–10	1,371
1,100–1,400	11–14	3,870
1,500–2,500	15–25	~25,100
2,600–10,000	26–100	~145,000
100–10,000	1–100	~175,000

The ratio of wanted to unwanted radiation is thus of the order of 5×10^{-6} and in order to reduce the unwanted radiation to less than 1 per cent, the filtering factor must be $\sim 10^7$.

As a further illustration, the results of a similar calculation performed by Lord and McCubbin (1957) are shown in Table II. It is evident from this table that when the first-order spectrum at 100 cm^{-1} is being observed, the band of second-order radiation near 200 cm^{-1} contains 7.5 times as much power as does the first-order band. The third-order band 300 cm^{-1} will contain 24 times as much power, and so on. It might be noted here that the values quoted in Table II represent the worst possible conditions, that is operation with a grating, equally efficient in all orders, at the centre of its blaze. However, since the angular width of the grating blaze is inversely proportional to the order number, operation slightly away from the blaze, while hardly affecting the first-order spectral intensity, will greatly reduce the intensity of the higher order spectra. In this sense, the echelette grating used away from its blaze position functions as a filter.

TABLE II. Factors Proportional to the Power in the Various Spectral Orders, Calculated for 2000°K Black Body Source Used in Conjunction with a Grating, Equally Efficient in All Orders.

Frequency (cm^{-1})	Spectral order					
	1	2	3	4	5	6
2000	1	3	3.2	2.7	1.5	0.88
1000	1	5.2	11.3	15.2	18.1	17.0
333	1	7.1	21	43	72	108
100	1	7.5	24	56	104	172
0	1	8	27	64	125	216

B. Types of Filter

The various filters employed to isolate "pure" first-order far-infrared radiation fall into four main categories: (1) absorption (or transmission) filters, (2) reflection filters, (3) scatter filters, and (4) interference filters. In general, a filter may act either as a "low-frequency pass filter", in that it will reject high-frequency radiation from the system; or as a "high-frequency pass filter", which rejects low-frequency radiation; or as a "narrow band pass filter", which isolates a narrow-frequency band of radiation. The basic requirement of a filter is that the transition from high to low transmission (or absorption or reflection) should be sharp—it is said that the filter should possess a sharp "cut-on" or "cut-off" at a particular frequency.

Classification into these four types may become somewhat arbitrary with regard to several of the filters discussed below. For example, it is customary to class black polyethylene as a transmission filter but in fact the modus operandi is scattering.

1. *Absorption or Transmission Filters*

(a) *Crystal Quartz*

A 1 mm thick wedge of crystal quartz is almost completely opaque between 2000–250 cm^{-1} and hence may be used as a low-frequency pass filter to isolate far-infrared radiation below 250 cm^{-1}. Fused quartz has its cut-on frequency at ~100 cm^{-1} and may be similarly used.

(b) *Black Polyethylene*

This consists of fine carbon particles dispersed in white polyethylene and is normally used in conjunction with crystal quartz to eliminate near-infrared wavelengths from the radiation beam. The small carbon particles preferentially scatter the shortwave radiation according to Rayleigh's Law (the scattering is proportional to $(\text{diameter})^3/(\text{wavelength})^4$). Filters (a) and (b) may be combined by using sooted quartz.

(c) *Polyethylene Embedded Reststrahlen Filters*

Yamada *et al.* (1962) observed that when powdered reststrahlen crystals were dispersed in polyethylene sheets at concentrations of some 10 per cent, complete absorption occurred at the reststrahlen (*vide infra*) frequency when the sheet thicknesses were about 3 mm. By suitable combination of crystals, they produced a series of such filters which, being opaque from some 800 cm^{-1} to their cut-on point (which varied from 400 cm^{-1} for a ZnO/BeO combination down to some 50 cm^{-1} for a TlCl/TlI/NaF combination), eliminate the radiation whose wavelengths lie immediately to the shortwave side of the region of interest. The higher-frequency radiation is, of course, removed by other types of filter.

The characteristics of these filters depend upon the particle size and concentration. To produce as sharp a cut-on as possible it is desirable to employ the lowest concentration and thinnest sample consistent with complete opacity in the reststrahlen region (optimum conditions appear to be 35% crystal and 3 mm thick sheets). The absorption due to scattering by the particles (and which is maximized at a wavelength of approximately half the particle diameter) should be made to reinforce the selective absorption of the crystals at their reststrahlen peak, by ensuring that the majority of the particles have a diameter of half the reststrahlen cut-off wavelength. A small fraction of the particles should have a diameter somewhat less than this value, in order to scatter the shortwave radiation.

(d) *Chopper—Modulation Filter*

Detectors employed in the far infrared are invariably designed to respond only to "chopped" (modulated or interrupted) radiation in order to facilitate amplification and the elimination of zero drift. If the chopper material is opaque only to far-infrared radiation, but transparent to higher frequencies, then the signal received by the detector will be due only to the low-frequency radiation. Commonly used materials are NaCl (opaque below 550 cm^{-1}), KBr (400 cm^{-1}) and CsI (225 cm^{-1}) (Blaine 1963).

The efficiency of this chopper modulation is not, however, complete since the chopper blades reflect some 4% of the higher-frequency radiation, at *each* surface. This 8% of "chopped" near-infrared radiation may well swamp the "chopped" far-infrared radiation, because the intensity of the near-infrared radiation is several orders of magnitude greater. The remedy is to block the open blades of the chopper with an opaque material, such as brass mesh, until the effect of reflection is compensated.

(e) *Semiconductor Filters*

Certain semiconductor materials (i.e. Si, Ge, InAs, InSb) act as very efficient low-frequency pass filters, since the absorption edges of these materials provide extremely sharp cut-ons. The transmission at frequencies below the cut-on may be considerably reduced due to reflection losses, as a result of the high refractive index of a semiconductor. Anti-reflecting layers are employed which enable some 80% transmission to be obtained in practice. There is no suitable material with a cut-on around 400 cm^{-1} and the main function of these filters is to reject completely all high frequency radiation above the cut-on frequency (~1500 cm^{-1} for InSb).

2. *Reflection Filters*

(a) *Reststrahlen Filters*

Certain ionic crystals have a high reflectivity, R, (0.90–0.95) over a relatively small frequency range in the far-infrared region—the position of peak reflection is termed the reststrahlen frequency (see Chapter 8). Away from this frequency the reflectivity falls to a low value (0.1), and hence the use of these crystals is an effective method of isolating far-infrared radiation of narrow bandwidth. A single reflection from a reststrahlen crystal can be seen to provide a rejection factor of some 10:1. In practice, two reflections are employed and it is found that (in conjunction with a compensated chopper and black polyethylene transmission filter) the spectral purity obtained is some 99%, especially if "crossed" reststrahlen crystals are used. If the angle of incidence is close to the Brewster angle then, after reflection, the reststrahlen reflected radiation will be unpolarized while the "unwanted" radiation will be polarized perpendicular to the plane of incidence. If the plane of incidence of the second reststrahlen crystal is arranged to be perpendicular to the plane of incidence of the first, then the effect on the "unwanted" radiation is analogous to the use of two crossed Nicol prisms in the visible region.

Turner *et al.* (1965) have indicated that the reflection characteristics of reststrahlen crystals may be improved by coating the filter with a nonabsorbing material. Ideally, reflection filters should have a wide, flat response and Turner demonstrated that the MgO reststrahlen band is made appreciably more rectangular by the use of a lead chloride coating.

(b) *Echelette Grating*

In addition to the use of an echelette grating as a dispersive element, a grating can also be used as a reflection filter. The use of a grating in this mode was first proposed by White (1947) to reject shortwave radiation. For radiation of wavelengths long compared with the groove separation, the grating will act as a mirror and specularly reflect the radiation. For radiation with wavelengths of the order of, or less than, the groove separation, the majority of the radiation goes into the order which causes the angle of diffraction to lie closest to the blaze angle. Hence the replacement of one of the plane mirrors in the optical path by an echelette grating greatly enhances the spectral purity of the low frequency radiation. Möller and co-workers (1965), for example, showed that when the high-frequency radiation intensity at a λ/d of less than 0.2 is reduced to zero, then the high-frequency radiation intensity at a λ/d of greater than unity is only reduced by 40 per cent. Alternatively, polyethylene replica gratings may be used in transmission, especially below 50 cm^{-1}.

3. *Scatter Filters*

Either wholly or partially, the action of black polyethylene, polyethylene embedded reststrahlen filters and any "rough surface" type of filter (echelette grating, roughened aluminium plate, wire meshes), depends upon the phenomenon of scattering.

(a) *Roughened Aluminium Plate*

Bennett and Porteus (1961) indicated that the ratio of the specular reflectance of a roughened surface, R_s, to the specular reflectance of a perfectly smooth surface, R_o, was proportional to $\exp(-(Ad/\lambda^2))$, where λ is the wavelength, d is the root mean square surface roughness, and A is a constant. Hence this form of filter operates in the low-frequency pass mode.

(b) *Reflection Meshes*

Following the use of a metal grid as a beam splitter in the Fabry-Perot interferometer (Renk and Genzel 1962), Yoshinaga and co-workers (1963) recommended its use as a reflection filter in the far infrared. They found experimentally that for five grids with inter-wire separations varying from 1/26–1/105 cm, the reflectance at $\lambda/d < 1.5$ was only some 5%, rising to some 95% at $\lambda/d > 2.1$.

Metallic grids may also be used in transmission, a particularly successful form being the transmission reciprocal mesh formed by depositing aluminium on to crystal quartz in such a way as to leave a transparent crystal quartz grid. Ressler and Möller (1967) showed that for a grid spacing of 1/200 cm, the transmission increased towards lower frequencies from its value of 5% at 100 cm^{-1}. As regards the isolation of pure first-order radiation, this form of filter is likely to become widely used below 60 cm^{-1} since at present no reststrahlen material is known with its peak below this frequency.

(c) *Christiansen Type Filters*

The Christiansen effect occurs with a suspension of particles in a transparent medium at a wavelength for which the refractive indices of the particle and medium are equal. At this wavelength the system is optically homogeneous and high transmission results. At other wavelengths the particles scatter the radiation, so reducing the transmission. Barnes and Bonner (1936) have given Christiansen transmission curves for the far-infrared region.

4. *Interference Filters*

These usually consist of several plane parallel layers of transparent materials with different refractive indices such that each interface between the materials has a finite reflectivity. They depend for their action on the interference of the multiply reflected beams within the system. Such devices may be designed to produce almost any desired spectral characteristic (reflection or transmission) by varying the number of layers.

A reflection-type interference filter simply consists of a thin layer of transparent dielectric deposited on a front surface mirror and a semi-reflecting metallic layer coated on the dielectric surface. A transmission interference filter consists of a thin layer of transparent dielectric material coated on each surface with a semi-reflecting metallic layer.

Metallic semi-reflecting layers absorb some 2% of the radiation at each interface, and when multiple reflections are involved this leads to an appreciable loss of transmission. If dielectric layers are used instead, then very little radiation (~0.2%) is lost at each interface. Filters may be produced, using such dielectric layers, with almost perfect transmission or reflection at an interface; these are called "all-dielectric interference filters".

The theoretical aspects of multilayer filters are somewhat complex, and apart from mentioning the result that such systems function as band pass filters with allowed (pass band) frequencies and forbidden (stop band) frequencies, the reader is referred to the appropriate literature (Hadley and Dennison 1947; Weinstein 1947; Caballero 1947; Tolansky 1948; Crook 1948; Smith 1958; Born and Wolf 1959.)

VI. Detectors

Problems of detection in the far infrared are considerable. An infrared detector converts the radiation incident upon it into an electrical signal which comprises, (1) the response to the radiant energy (the signal), and (2) the inherent fluctuations of the detector ("noise"). The usefulness of a detector will be limited by the relative magnitudes of the signal and the noise and, in a far-infrared grating spectrometer, difficulties associated with the low intensity of the required radiation emitted from the source are aggravated by the fact that only a single resolution width is incident upon the detector at a given instant, thus reducing the signal-to-noise ratio.

Practical detectors may be divided into two types depending upon whether the heating effect of the incident radiation is observed (thermal detectors) or whether the interaction of the incident radiation with the detector material, freeing bound electrons (photo-detectors) is studied.

The main difference between the two types of detector is that, whereas a thermal detector responds to the total energy absorbed, the photo-detector only responds to "effective" quanta of radiation. An effective quantum must have an energy not less than an energy E_0, characteristic of the detector. Quanta with energy less than E_0 are not "counted" by the photo-detector, which thus has an associated long-wave limit of detection, given by $\lambda = hc/E_0$. The thermal detector, on the other hand, responds equally to each and all wavelengths.

Detectors may be characterized by four quantities:

(1) The spectral or wavelength response—this is usually peaked and limited for photo-detectors but "flat" for thermal detectors.

(2) The response time—for the observation of transient phenomena very fast response times (10^{-7} sec) are required but for absorption measurements in the far infrared, there is no particular advantage in using a fast detector.

(3) The responsivity—this determines the magnitude of the response to a given radiation flux.

(4) The ultimate sensitivity—this determines the minimum radiation flux observable with the detector under given conditions.

When trying to compare the relative performances of the various detectors, two additional quantities arise. These are (a) the minimum detectable power, W_m, which is the incident power which will give a signal-to-noise ratio of unity at the output of the detector, and (b) the noise equivalent power, NEP, which is given by the minimum detectable power divided by the square root of the amplifier band width, $W_m/\Delta f^{1/2}$.

The numerical index D^*, called the detectivity, was suggested by Jones (1952) as an indication of the relative merit of a detector. The detectivity is the reciprocal of the noise equivalent power for a detector having unit area and used with an amplifier with a 1 c/sec band width: $D^* = (S/N)(A\Delta f)^{1/2}/W$ where (S/N) is the signal-to-noise ratio obtained when W watts of radiation are incident on the detector of area A cm^2, used in conjunction with an amplifier of bandwidth Δf (c/sec).

On this basis it appears that there is little to choose between the superconducting bolometer, the boron or antimony-doped germanium photoconductor and the indium antimonide detector, which all possess a four-fold higher detectivity than the carbon bolometer (Richards 1965). Putley (1965) and Martin and co-workers (1961) demonstrated a forty-fold superiority in the signal-to-noise ratio of a superconducting bolometer over a Golay detector.

A. Thermal Detectors

Although several thermal detectors have been used in the past, the only ones to merit attention here are the cooled bolometers and the Golay pneumatic detector.

1. *Cooled Bolometers*†

In this type of detector a small resistance thermometer measures the temperature change in the receiving element. The noise consists of Johnson noise (fluctuations in the voltage across the resistor) and thermal noise (fluctuations in the temperature of the detecting element due to the statistical nature of thermal equilibrium), both of which increase rapidly with decreasing temperature. Unlike ordinary metal bolometers, in which the temperature coefficient also decreases with temperature and thus the reduction in noise is effectively lost, superconducting metal bolometers possess very high-temperature coefficients of resistance and very low noise levels and function as high-detectivity detectors. In practice, the detecting element has been constructed either

† How can one describe a bolometer?
It's a sensitive kind of thermometer
Which will measure the heat
From a polar bear's seat
At a distance of half a kilometre.

from tin (Martin *et. al.* 1961), carbon (Boyle and Rodgers 1959) or doped germanium (Low 1961).

2. *Golay Pneumatic Detector*

The Golay cell (Golay 1947, 1949) employs a small constant-volume gas thermometer to measure the changes in temperature of a thin foil receiving element. The absorption of radiation increases the pressure in the thermometer with the result that a thin membrane, which forms one wall of the thermometer, is distorted. The signal is amplified by reflecting a light beam off the membrane into a photocell amplifier.

This detector is widely used in the far infrared for the following reasons: (a) it may be used in a vacuum, (b) due to the relatively large window area, the focussing and direction of the incident radiation is not critical, (c) it does not require cooling, and (d) it is commercially available and fairly robust. These factors far outweigh the slight loss in detectivity when compared with other detectors.

B. Photo-detectors

These detectors depend for their operation on the effects resulting from the interaction of the incident radiation with the electrons in the detector material. Four main effects have been distinguished (the photoemissive, photovoltaic, photoconductive, and photo-electromagnetic effects) but only the photoconductive effect in semiconductor materials is of importance for far-infrared detectors. In such materials, electrons in bound states in the valence band absorb the incident radiant energy and are promoted into free states in the conduction band. Electrical conduction may result either from electrons in the conduction band or from positive holes in the valence band.

Two types of photoconductor may be recognized, depending upon whether the promotion occurs across the energy gap between the valence and conduction bands of the semiconductor (intrinsic photo-conduction) or between intermediate energy levels due to the presence of impurities (extrinsic photo-conduction). The low-frequency threshold of detection is governed by the effective energy gap ΔE and is given by $h\nu = \Delta E$ or

$$\nu\ (\text{cm}^{-1}) = \Delta E\ (\text{eV})\ 10^4/1.23.$$

Almost all intrinsic semiconductors either have an energy gap greater than 0.2 eV, corresponding to threshold frequencies in excess of 800 cm^{-1}, or they have negligibly small energy gaps. All practical photodetectors operating in the far infrared are thus of the extrinsic type.

Typical examples of extrinsic semiconductor far-infrared detectors are the boron-doped germanium detector (Shenker *et al.* 1965) and the antimony-doped germanium detector developed by Fray and Oliver (1959), both of which operate around 100 cm^{-1}. For lower frequencies the indium–antimonide detector is used which has a working range of 1–50 cm^{-1}; this depends on the photo-conductivity produced in extremely pure, cooled InSb when placed in a magnetic field of some 3–10 kgauss.

VII. Interferometry

A. Introduction

Apart from improvements in conventional prism-grating instrumentation, recent research in far-infrared spectroscopy has been successfully applied to a different technique, namely interferometry. Interferometers of several types, notably the Fabry-Perot, lamellar grating and Michelson varieties, have been adapted by various workers for absorption measurements, but the type most commonly used by chemists to date is the Michelson interferometer.

All interferometers essentially conform to the following description of a Michelson variety. Radiation from a far-infrared source (usually a mercury lamp) is chopped, rendered parallel and then divided by a beam splitter (which is a lamellar grating in the lamellar grating interferometer, either a metal mesh or taut film of polyethylene terephthalate ("Mylar", "Melinex") in the Michelson interferometer, and a pair of parallel plates of high reflectivity and low absorption in the Fabry-Perot interferometer). Part of the radiation is reflected on to a movable mirror and part is transmitted to a fixed mirror. The radiation travels back from the mirrors along its incident paths and is recombined at the beam splitter, whence it passes through the sample to the detector (usually a Golay cell). The amplified signal is simultaneously recorded and digitized on paper tape (or cards) for input to a computer. The trace of signal against path difference is known as an "interferogram", the Fourier transform of which has long been known to yield the absorption spectrum of the sample (Michelson 1902).

The movable mirror introduces a variable path difference between the beams before recombination. When the two beams are of equal length, all the frequencies are in phase and a maximum output is obtained at the detector—this situation is referred to as "zero path difference". When the two optical paths are unequal, the different frequencies are no longer in phase, with the result that some interfere constructively and others destructively, producing a characteristic curve.

B. Theory

The interferogram obtained from radiation of the power spectrum $S(k)$ with respect to wavenumber k may be represented as a function of path difference x by the general Fourier transform

$$I(x) = \int_{-\infty}^{\infty} S(k) \exp(-2\pi ikx)\, dk \tag{14}$$

and hence the power spectrum is given by

$$S(k) = \int_{-\infty}^{\infty} I(x) \exp(2\pi ikx)\, dx. \tag{15}$$

If the interferogram is symmetrical about zero path difference then the sine transform part of $\exp(2\pi ikx) \equiv \cos 2\pi kx + i \sin 2\pi kx$ disappears, resulting in the expressions

$$I(x) = 2\int_{0}^{\infty} S(k) \cos 2\pi kx\, dk \tag{16}$$

and

$$S(k) = 2\int_{0}^{\infty} I(x) \cos 2\pi kx\, dx. \tag{17}$$

For the sake of simplicity the symmetrical interferogram case will be considered. Equation (17) represents the Fourier transformation which must be carried out in order to evaluate the spectrum from the measured interferogram.

Computational errors must arise since the integral in eqn (17) cannot be evaluated exactly. Numerical integration techniques must be employed and the most appropriate method in this case is the trapezoidal rule. The computed spectrum $S(k)$ will then have the form

$$\bar{S}(k) = q\{I(0) + 2I(q) \cos 2\pi kq + \ldots + I(nq) \cos 2\pi k . nq\} \tag{18}$$

where $I(nq)$ is the interferogram value at a path difference nq and q is the difference between consecutive values of x at which the interferogram is sampled and is termed the sampling period.

The above approximation (eqn (17) to eqn (18)) results in two distinct sources of error: (1) due to the use of a finite sampling period q instead of zero, and (2) due to truncating the integral from infinity to nq. These will be discussed separately below.

1. *Sampling Error*

If eqn (18) were replaced by an infinite series

$$\bar{S}_s(k) - q\{I(0) + 2I(q) \cos 2\pi kq + 2I(2q) \cos 2\pi k . 2q + \ldots\} \tag{19}$$

then no error would arise due to finiteness of the sampling period. $\bar{S}_s(k)$ is the exact cosine Fourier transform of the function $I_s(x)$, which is obtained by multiplying the interferogram function $I(x)$ by a Dirac comb function:

$$\bar{S}_s(k) = 2\int_0^\infty I_s(x)\cos 2\pi kx\,dx.... \tag{20}$$

By the convolution theorem (Blackman and Tukey 1958)

$$\bar{S}_s(k) = S(k) * D'$$

where D' is the transform of the Dirac comb function D. It may be shown that

$$D' = \ldots + \delta\left(k - \frac{2}{q}\right) + \delta\left(k - \frac{1}{q}\right) + \delta(k) + \delta\left(k + \frac{1}{q}\right) + \delta\left(k + \frac{2}{q}\right) + \ldots \tag{21}$$

and hence

$$\bar{S}_s(k) = \ldots + S\left(k - \frac{1}{q}\right) + S(k) + S\left(k + \frac{1}{q}\right) + \ldots \tag{22}$$

The individual terms in the above infinite series are called *aliases* and the displacement between consecutive aliases $(1/q)$ is termed the aliasing period.

It can be shown that if $S(k)$ is zero for $|k|$ greater than a given frequency $k_0 = 1/2q$ then

$$\bar{S}_s(k) = S(k) \quad \text{for} \quad |k| < \frac{1}{2q}.$$

Thus if $S(k)$ vanishes outside the principal aliasing interval

$$-\frac{1}{2q} < k < \frac{1}{2q}$$

then the computed spectrum $\bar{S}_s(k)$ is identical with the true spectrum $S(k)$. Practically this is achieved by filtering out all frequencies greater than $k_0 = 1/2q$.

2. *Truncation Error*

The effect of only being able to sample the interferogram to a maximum path difference nq instead of to infinity may be represented as

$$\bar{S}_T(k) = 2\int_0^{nq} I(x)\cos 2\pi kx\,dx \tag{23}$$

and this may be written as

$$\bar{S}_T(k) = 2\int_0^\infty I_T(x)\cos 2\pi kx\,dx. \tag{24}$$

$\bar{S}_T(k)$ is the exact cosine Fourier transform of the function $I_T(x)$, obtained by multiplying $I(x)$ by a crenel function, C, which is unity for $|x| < nq$ and zero for $|x| > nq$.

The convolution theorem gives

$$\bar{S}_T(k) = S(k) * C'\left(k; \frac{1}{nq}\right)$$

where $C'(k; 1/nq)$ is the transform of the crenel function, and is often referred to as the crenel-mate. The properties of this function are such that any fine structure in the spectrum $S(k)$, comparable in magnitude with $1/2nq$, will be blurred out. The lateral extent of blurring is characterized by a limit of resolution, $1/2nq$, which is inversely proportional to the truncation length, $2nq$. This is because, although the crenel-mate has a grand maximum at the origin, zeros at $\pm m/2nq$ ($m = 1, 2, 3$, etc.) and turning points, alternatively minima and maxima, at $\pm l/4nq$ ($l = 3, 5, 7$, etc.) for most practical purposes the function may be considered negligible outside the region $\pm 1/2nq$ from the origin.

However, the truncation errors may be reduced by using a function similar to the crenel-mate, but with less prominent "side-lobes". This is achieved by introducing a weighting function $\omega(x/nq)$, whose transform is $\Omega(k; 1/nq)$. Now $\bar{S}_T(k)$ is given by

$$\bar{S}_T(k) = S(k) * C'\left(k; \frac{1}{nq}\right) * \Omega\left(k; \frac{1}{nq}\right)$$

in which the function, with which $S(k)$ is convolved, is known as the "apodized spectral window" because if a suitable weighting function is chosen, the effect of convolving C' with Ω is to reduce the size of the side lobes of C'.

The choice of weighting function is arbitrary to a large degree but it may be shown that the more structure appearing in $S(k)$, the more sophisticated a weighting function will be required to reduce errors due to truncation. This increased control of truncation error by apodization is however achieved only at the extent of resolution: the resolution limit is given by $\Delta k = (p + 1)/2nq$ where p is the degree of sophistication of the weighting function. Hence in choosing the weighting function, the limit of resolution is matched to the fineness of detail which has to be resolved in the spectrum. It may be necessary, if the spectral structure is very fine, to compute an unapodized spectrum, $p = 0$: $\Delta k = 1/2nq$. A commonly used weighting function has the form: $(p = 1)$ $\omega(x/nq) = \cos^2(\pi x/2nq)$, which has a resolution limit $\Delta k = 1/nq$, demonstrating the loss of resolution due to the increased apodization efficiency.

Thus it has been shown that, (1) to avoid aliasing errors it is necessary to exclude radiation of frequency greater than $k_{\max} = 1/2q$ by the use of filters, and (2) truncation errors may be minimized by the use of a suitable weighting function. Using the weighting function $\cos^2(\pi x/2nq)$, eqn (18) may now be written as

$$\bar{S}(k) = q\left\{I(0) + 2I(q)\cos^2\left(\frac{\pi}{2}\cdot\frac{1}{n}\right)\cos 2\pi kq + \ldots \right.$$
$$\left. + 2I((n-1)\,q)\cos^2\left(\frac{\pi}{2}\frac{n-1}{n}\right)\cos 2\pi k(n-1)\,q\right\} \quad (25)$$

which is the basis of any digital computer program for numerical Fourier transformation.

Application of information theory (Goldman 1953) provides the result that in order to obtain all the information in a spectrum between 0 and $k_{\max}$ it is necessary to sample the interferogram at intervals of $\Delta x \leqslant \frac{1}{2}k_{\max}$.

If the interferogram is not sampled at zero path difference and equally spaced points $\Delta x = q$, $2q$, $3q$ etc., but at a position close to zero path difference $\epsilon(-q/2 < \epsilon < q/2)$ and equally spaced points $\Delta x = (q+\epsilon)$, $(2q+\epsilon)$ etc., then mere replacement of the interferogram ordinates $I(0), I(q), I(2q)\ldots I(nq)$ by $I(\epsilon), I(q+\epsilon), I(2q+\epsilon)\ldots I(nq+\epsilon)$ in eqn (25) will result in large errors in the computed spectrum.

Gibbs and Gebbie (1965) have shown that by modifying the computation formula to that given below, these errors are eliminated:

$$\bar{S}(k) = q\left\{I(\epsilon)\cos^2\left(\frac{\pi}{2}\frac{\epsilon}{nq}\right)\cos 2\pi k\epsilon + 2I(q+\epsilon)\cos^2\left(\frac{\pi}{2}\frac{q+\epsilon}{nq}\right)\cos 2\pi k(q+\epsilon)\right.$$
$$\left. + \ldots + I(nq+\epsilon)\cos^2\left(\frac{\pi}{2}\cdot\frac{nq+\epsilon}{nq}\right)\cos 2\pi k(nq+\epsilon)\right\} \quad (26)$$

To illustrate the practical implications of this section, a short example follows. If it is desired to investigate a spectrum over the frequency range 0–200 cm^{-1} with a resolution, Δk, of 0.2 cm^{-1}, then the total mirror travel must be 5 cm; all frequencies above 200 cm^{-1} must be filtered out; the sampling of the interferogram must occur at least every 25 μ ($=\frac{1}{2}(200\text{ cm}^{-1})$) of path difference and the number of sampling points equivalent to this would be $n = 2k_{\max}/\Delta k = 2(200)/0.2 = 2000$.

C. Advantages of Interferometry

This type of spectrometer has several advantages over a grating monochromator. With an interferometer the whole spectral range of interest is simultaneously incident upon the detector at any given time, rather than merely a single resolution width as with a grating instrument. This implies an increase in the signal-to-noise ratio, and is usually known as the Fellgett advantage (Fellgett 1959). Since, at very low frequencies, the signal-to-noise ratio is a limiting factor, this is an important advantage for measurements at these frequencies. The greater energy throughput also means that it is easier to examine small samples if necessary, for example in a diamond-windowed high pressure cell. Secondly, since the resolution of an interferometer is proportional to the maximum path difference (twice the total distance travelled by the movable mirror), in order to double the resolution of a given spectrum, the mirror travel, and hence the measuring time, is doubled. In the case of a grating instrument, both the entrance and exit slits must be halved, so reducing the energy four-fold. To obtain a spectrum with double the resolution and the same signal-to-noise ratio, the measuring time must be increased sixteen-fold. A third advantage is that the filtering system required with an interferometer is much simpler than the complex system employed with a grating instrument.

The main disadvantage of Fourier transform spectroscopy is the possible delay in gaining access to facilities with which to compute the spectrum. In this sense it is true that the spectrum is not measured directly, and therefore optimization of sample thickness or the effect of varying a parameter, for example, is more difficult. By the same token it is more difficult to recognize unsatisfactory conditions during a run, for example, the decomposition of the sample. It should also be remembered that the occurrence of a single noise-spike in the interferogram will affect *all* the frequencies in the spectrum. However, the use of a computer to ratio spectra and evaluate absorption intensities must be considered to offset this time delay, which is gradually disappearing with the advent of small computers designed to work in conjunction with an interferometer.

D. Modes of Operation of an Interferometer

An interferometer is said to be used in either the "aperiodic" or "periodic" mode depending upon whether the path difference is varied uniformly with time without repetition or whether it is varied periodically with a constant velocity saw-tooth motion, respectively (see Fig. 2).

When used in the periodic mode, the path difference of the interferometer is varied with a saw-tooth motion of period $2T_0$ (Fig. 2), which is usually of the order of ten seconds. This results in each resolution width of the spectrum appearing as an audio frequency harmonic of the frequency $k_0 = 1/T_0$. The spectrum is produced by feeding the signal from the phase-sensitive detector through an amplifier, of bandwidth less than k_0, tuned in turn to each harmonic.

This method does not utilize the Fellgett advantage since only a single harmonic at a time is sampled. However, if the signal is fed through a series of amplifiers, each tuned to a separate harmonic, then the Fellgett advantage is restored. Considerable technical difficulties are obviously associated with the periodic use of an interferometer, the major one being the production of the periodic drive mechanism. Such difficulties have set upper limits on both the resolution and operating range of practical periodic instruments.

When an interferometer is used in the aperiodic mode, the interferogram is sampled at small uniform intervals of path difference in a single

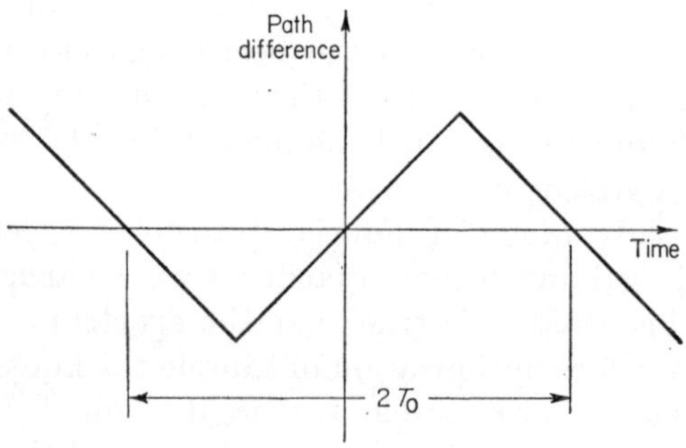

FIG. 2. Periodic path difference change for an interferometer.

slow sweep. The interferometer is driven rapidly between sampling points, where it is stopped long enough to average the noise to an acceptable value. The spectrum is evaluated using a high speed digital computer.

An aperiodic interferometer is simpler to construct, may be used over a wider spectral range and also, since the attainable maximum path difference is greater, at a higher resolution than a periodic instrument. For these reasons the most commonly used interferometers are of the aperiodic type with the Michelson variety enjoying more popularity than the lamellar grating.

E. Choice of Interferometer

The choice of actual instrument depends on the exact frequency range of interest, the quantity of real concern being the beam splitter efficiency. An ideal beam splitter would have transmission and reflection coefficients, T and R respectively, both equal to a half. The ratio of the product of the experimentally determined values of T and R, to the theoretical product, 0.25, gives the beam splitter efficiency.

Richards (1964) determined the beam splitter efficiency of a 3-mil Mylar beam splitter and of a lamellar grating (grating constant 1.9 cm). The results are shown in Fig. 3.

It can be seen that the lamellar grating has an efficiency near 100% for frequencies below a certain frequency, which is inversely proportional to the grating constant, but that the efficiency gradually falls towards higher frequencies. Decreasing the grating constant in order to increase the frequency below which the efficiency becomes near ideal, unfortunately also decreases the resolution.

The Mylar beam splitter can be seen to be most useful in the region of its first interference maximum. In practice, the beam splitter thickness

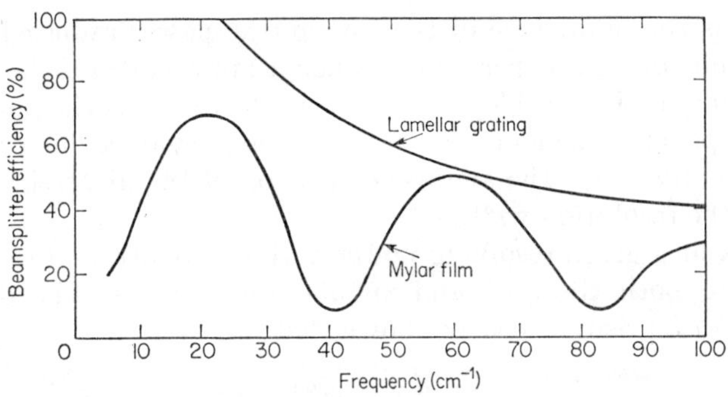

FIG. 3. Comparison of the beam splitter efficiencies of a lamellar grating (grating constant $d = 1.9$ cm) and a mylar film (3 mil).

is selected such that k_{max}, the highest frequency of interest in the spectrum, coincides with the first interference minimum, thus helping the filtering. A thinner beam splitter will increase the relative energy available at higher frequencies, and will also increase the usable frequency range of a single beam splitter. However, the beam splitter efficiency falls off very rapidly at frequencies below some $0.2K'$, where K' is the frequency of the first interference maximum. Hence several beam splitter changes would be required to cover the frequency range below (say) 20 cm^{-1}.

Below about 20 cm^{-1}, therefore, it is desirable to use a lamellar grating interferometer, with its inherent beam splitting advantage. Above this frequency, despite the fact that the lamellar grating is still superior to the Michelson interferometer, it is not as widely used, due to the relative ease of construction of the latter instrument.

VIII. Comparison of Prism, Grating and Interferometric Spectrometers

The two most important characteristics of a monochromator or spectrometer are (1) its ability to resolve closely spaced spectral components which is characterized by the resolving power $R = \lambda/d\lambda = k/dk$ (λ = wavelength, k = frequency), and (2) the luminosity or flux collected by the device from a source emitting unit power per unit area per unit wavelength interval. This obviously depends upon the resolving power of the device and hence any comparison in luminosity between spectroscopic instruments should strictly be made at the same resolving power.

For slit instruments (that is the prism and grating spectrometers) the flux passed by the output slit is given by

$$F = B_\lambda \frac{l}{f} A T \frac{d\alpha}{d\lambda} (\Delta\lambda)^2$$

where B_λ is the luminance of the source (the power radiated per unit area per unit solid angle per unit wavelength internal); l is the length of the exit slit; f is the focal length of the collimator; A is the normal area of the dispersing element; T is the transmission coefficient of the instrument; $(d\alpha/d\lambda)$ is the angular dispersion of the dispersing element and $\Delta\lambda$ is the resolution width.

The flux at a given resolving power is thus proportional to the ratio (l/f) and to both the area and angular dispersion of the dispersing element. For a prism it may be shown that

$$A \frac{d\alpha}{d\lambda} = A' \frac{dn}{d\lambda}$$

where A is the base area of the prism, whose refractive index is n. For a grating the product of normal area, A_n, and angular dispersion is given by

$$A_n \frac{d\alpha}{d\lambda} = \frac{2A \sin\beta}{\lambda}$$

where β is the grating blaze angle.

The relative performance of a grating instrument compared with a prism instrument, at constant normal area and resolving power is thus seen to be

$$F_P/F_G = \lambda(dn/d\lambda)/2\sin\beta.$$

The blaze angle of a grating is typically 30° yielding $F_P/F_G = \lambda(dn/d\lambda)$. This means that the comparison depends wholly on the value of the dispersion of the prism material. Graphs of $\lambda(dn/d\lambda)$ against λ for practical far-infrared prism materials indicate that the grating has approximately a ten-fold advantage over the prism. In addition to this, gratings may be manufactured with larger areas than prisms and this again favours the grating instrument.

The flux passed by an interferometric device can be shown to be approximately given by

$$F_I = 3.4 B_\lambda T A(\Delta\lambda)^2/\lambda$$

where the various symbols have been defined previously.

The ratio of fluxes passed by the interferometric and grating spectrometers is thus given by

$$F_I/F_G = \frac{3.4 B_\lambda T A(\Delta\lambda)^2/\lambda}{B_\lambda(l/f)\, 2A \sin\beta T(\Delta\lambda)^2/\lambda} = \frac{3.4}{(l/f)} \qquad \text{for } \beta = 30^\circ.$$

Practical grating spectrometers have a typical (l/f) value of the order of 1/50. This implies a flux gain of some two orders of magnitude in favour of the interferometer over the grating device.

It is thus apparent that, as far as flux passed is considered for equivalent dimensioned devices at a given resolution, the interferometer is vastly superior to the grating spectrometer, which in turn has a marked superiority over the prism spectrometer. A detailed discussion of the relative merits of interferometry and grating spectroscopy, together with pertinent data, forms the subject of the several papers in the March 1969 issue of *Applied Optics* (**8**, 497–705).

IX. Commercial Far-Infrared Spectrometers

Prior to this decade most far-infrared spectrometers were "home made" and their operation could not always be described as routine. At the present time there is a wide range of commercially available spectrometers to cover the region down to 200 cm^{-1}, and a more limited choice of specialized grating instruments and interferometers extending to considerably lower frequencies.

A. Grating Spectrometers

It is pointless to attempt to give a comprehensive list of available instruments as this would rapidly become out of date. Some of the currently available models, however, together with their frequency ranges, are listed in Table III.

These spectrometers fall into two broad categories: (1) basic mid-infrared spectrometers which have had their range extended downwards from 400 cm^{-1} to the 200–300 cm^{-1} region, and (2) those which are designed specifically as far-infrared instruments.

TABLE III. Some Commercially Available Grating Spectrometers Covering the Far-Infrared Region.

Manufacturer	Model	Frequency range (cm^{-1})
Beckman	IR 11	800–33 cm^{-1}
Beckman	IR 12	4000–200 cm^{-1}
Beckman	IR 4	Various optics available down to 285 cm^{-1}
Beckman	IR 20A	4000–250 cm^{-1}
Grubb Parsons	GM 3	200–65 cm^{-1}
Grubb Parsons	DM 4	667–222 cm^{-1}
Grubb Parsons	DM 4	500–200 cm^{-1}
Hitachi	EPI-L	700–200 cm^{-1}
Hitachi	FIS-3	500–60 cm^{-1}
Perkin Elmer	301	667–14 cm^{-1}
Perkin Elmer	225	5000–200 cm^{-1}
Perkin Elmer	621	4000–200 cm^{-1}
Perkin Elmer	521	4000–250 cm^{-1}
Perkin Elmer	421	4000–250 cm^{-1}
Perkin Elmer	457	4000–250 cm^{-1}

In principle, it is not difficult to extend the range of mid-infrared spectrometers to 200 cm^{-1} as standard sources (for example Nernst and Globar) and detectors (provided that they are equipped with a suitable transparent window) are still effective at these frequencies. Thus a number of basic mid-infrared spectrometers are now equipped to measure to 200 cm^{-1}, two examples of these being the Beckman IR 12 and the Perkin Elmer 225. Beyond 200 cm^{-1}, as discussed earlier in this chapter, other design considerations become necessary and the complexity of the instrumentation increases. Among spectrometers designed to cover these lower frequencies are the Perkin Elmer 301, the Beckman IR 11, Grubb Parsons GM 3 and Hitachi FIS-3. Some of the chief characteristics of these spectrometers are listed in Table IV.

As an example of one modern grating spectrometer, the optical diagram of the Beckman IR 11 is reproduced below (Fig. 4, see also Table IV). Double beam optical null operation is employed over the entire range (800–33 cm^{-1}). Four gratings (each used only in the first order and near its blaze) cover this range and are mounted on a rotatable turret. Crystal choppers and reststrahlen filters are not used and the elimination of stray light is accomplished by transmission filters which are also mounted on a turret. External switches control the selection of gratings and transmission filters. Removal of water vapour is accomplished with a special air-drying unit. As a demonstration of the ease of operation, a complete scan from 800 to 33 cm^{-1} can be made in as little as nine minutes.

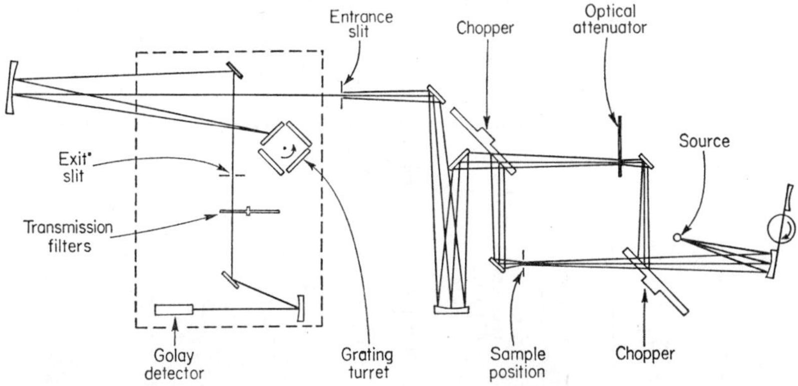

FIG. 4. Optical diagram for the Beckman IR 11 far-infrared spectrometer.

B. Interferometers

Commercial models are produced by Grubb Parsons and by Research and Industrial Instruments Company (R.I.I.C.). The interferometer systems are usually considerably cheaper than the grating spectrometers which extend to the 15–30 cm^{-1} region. However, it must be remembered that if the cost of a computer to carry out the Fourier transform is included, the price rises sharply.

All the instruments are operated under vacuum and are mostly modular in construction. The great advantage of the modular construction is the possibility for interchange of various units and the ease with which various sample handling facilities can be incorporated (for example low temperature cells, long path gas cells). With the exception of the LR 100 all the instruments are of the Michelson type and each has alternative thicknesses of beam splitter for maximum efficiency over the various low-frequency ranges.

Apart from the basic electronics, R.I.I.C. market a further unit (FTC 300) which performs the Fourier transform on the interferogram. This allows the presentation of the required spectrum on a recorder shortly after the completion of the interferogram and can also ratio sample against background spectra in order to remove background features.

In Table V the main characteristics of the commercially available interferometers are listed. Examples of (1) a Michelson and (2) a lamellar grating instrument are now briefly described.

1. *Michelson Type*

As an example of a commercial Michelson interferometer, the optical diagram for the Grubb Parsons—NPL "Cube" instrument is shown in

TABLE IV

Spectrometer	Range	Mode of operation	Monochromator	Source	Detector	Resolution	Stray light	Water vapour removal
Beckman IR 11	800–33 cm^{-1}	Double beam or single beam	4 gratings 3 l/mm blazed at 165 μ 8 l/mm blazed at 112 μ 20 l/mm blazed at 45 μ 50 l/mm blazed at 19.98 μ	High pressure Hg lamp	Golay (diamond window)	0.5–1.0 cm^{-1} over most of the range	Less than 4% at 80 cm^{-1}	Dry air purge
Grubb-Parsons GM3	200–65 cm^{-1}	Single beam	1 grating 8 l/mm blazed at 110 μ	High pressure Hg lamp	Golay (quartz window)	Better than 3 cm^{-1}	Less than 2% at 70 cm^{-1}	Dry air purge
Hitachi FIS-3	400–33 cm^{-1}	Double beam or single beam	3 gratings 4 l/mm 8 l/mm 20 l/mm	Globar and high pressure Hg lamp	Golay (diamond window)	0.5 cm^{-1}	—	Evacuated
Perkin Elmer 301	667–14 cm^{-1}	Double beam to 32 cm^{-1}. Single beam 32–14 cm^{-1}	5 gratings 2.5 l/mm blazed at 360 μ 5 l/mm blazed at 180 μ 10 l/mm blazed at 90 μ 20 l/mm blazed at 45 μ 40 l/mm blazed at 22.5 μ	Globar and high pressure Hg lamp	Golay (diamond window)	0.7 cm^{-1}	—	Dry air purge

TABLE V

Manufacturer	Model	Type	Range	Optical system	Source	Detector	Maximum resolution
Block Engineering	FTS 16	Michelson (modular)	650–5 cm^{-1}	Polyethylene terephthalate beam splitter; lenses for condensing optics	High pressure Hg arc	Golay	0.1 cm^{-1}
Grubb Parsons	IS 3	Michelson	675–10 cm^{-1}	Polyethylene terephthalate beam splitter; all reflecting optics	High pressure Hg lamp	Golay	0.1 cm^{-1}
Grubb Parsons	NPL "Cube"	Michelson (modular)	675–10 cm^{-1}	Polyethylene terephthalate beam splitter; polyethylene lens for condensing optics	High pressure Hg lamp	Golay	0.2 cm^{-1} (0.1 cm^{-1} with long travel drive unit)
R.I.I.C.	FS 820	Michelson (modular)	200–10 cm^{-1}	Polyethylene terephthalate beam splitter; polyethylene lens for condensing optics	High pressure Hg lamp	Golay	0.2 cm^{-1}
R.I.I.C.	FS 720	Michelson (modular)	500–10 cm^{-1}	Polyethylene terephthalate beam splitter; all reflecting optics	High pressure Hg lamp	Golay	0.1 cm^{-1}
R.I.I.C.	LR 100	Lamellar grating (modular)	80–3 cm^{-1}	Lamellar grating and reflecting optics; polypropylene condensing lens	High pressure Hg lamp	Golay	0.1 cm^{-1}

Quartz windows are usually employed if only the range 5–200 cm^{-1} is required; diamond windows are necessary for higher wave-number ranges although silicon is now also employed.

Fig. 5. The radiation is chopped and passes directly to the beam splitter (polyethylene terephthalate) and then to the moving and fixed mirrors.

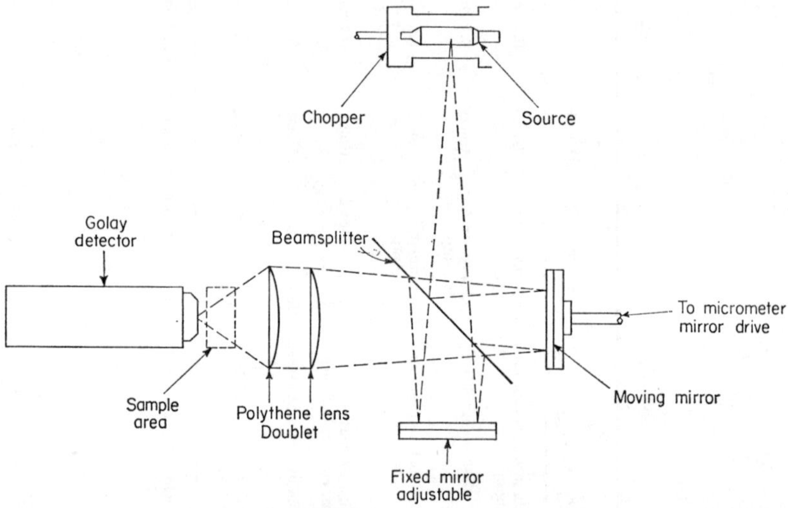

Fig. 5. Optical diagram for the Grubb Parsons NPL "Cube" interferometer.

After recombination of the beams the light passes to a polyethylene lens doublet which condenses the beam to the sample chamber and thence the Golay detector.

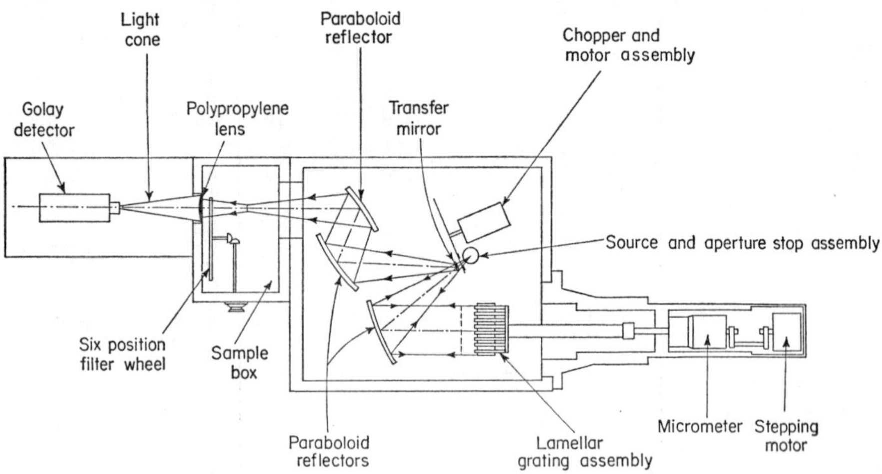

Fig. 6. Optical diagram for the RIIC LR 100 lamellar grating interferometer.

2. *Lamellar Grating Type*

The R.I.I.C., LR 100 lamellar grating interferometer is illustrated in Fig. 6. The lamellar grating itself is 8 × 8 cm and comprises two sets of parallel intermeshing metal plates whose front surfaces have been lapped accurately flat. Radiation from the source is chopped and then collimated by a surface-aluminized off-axis parabolic mirror whence it passes to the lamellar grating. Variation of the path difference is effected by a stepping motor coupled to the micrometer drive of the lamellar grating. Radiation reflected by the grating is deflected by the off-axis parabolic mirror to a transfer mirror directly beneath the source aperture from where it passes, *via* two off-axis parabolic mirrors, to the sample chamber and detector.

References

Barnes, R. B. and Bonner, L. G. (1936). *Phys. Rev.* **49**, 732.

Benesch, W. and Strong, J. (1951). *J. opt. Soc. Am.* **41**, 252.

Bennett, H. E. and Porteus, J. O. (1961). *J. opt. Soc. Am.* **51**, 123.

Blackman, R. B. and Tukey, J. W. (1958). "The Measurement of Power Spectra." Dover Publications, New York.

Blaine, L. R. (1963). *J. Res. natn. Bur. Stand.* **67c**, 207.

Born, M. and Wolf, E. (1959). "Principles of Optics" Pergamon Press, Oxford.

Boyle, W. S. and Rogers, K. F. (1959). *J. opt. Soc. Am.* **49**, 66.

Caballero, D. (1947). *J. opt. Soc. Am.* **37**, 176.

Cano, R. and Mattioli, M. (1967). *Infrared Phys.* **7**, 25.

Crook, A. W. (1948). *J. opt. Soc. Am.* **38**, 954.

Czerny, M. and Turner, A. F. (1930). *Z. Phys.* **61**, 792.

Fastie, W. G. (1952). *J. opt. Soc. Am.* **42**, 641.

Fellgett, P. (1958). *J. Phys. Radium, Paris* **19**, 187, 273.

Fray, S. J. and Oliver, J. F. C. (1959). *J. scient. Instrum.* **36**, 195.

Genzel, L. and Eckhardt, W. (1954). *Z. Phys.* **139**, 578.

Gibbs, J. E. and Gebbie, H. A. (1965). *Infrared Phys.* **5**, 187.

Golay, M. J. E. (1947). *Rev. scient. Instrum.* **18**, 357.

Golay, M. J. E. (1949). *Rev. scient. Instrum.* **20**, 816.

Goldman, S. (1953). "Information theory." Prentice Hall, Englewood Cliffs, New Jersey.

Hadley, L. N. and Dennison, D. M. (1947). *J. opt. Soc. Am.* **37**, 451.

Jones, R. C. (1952). *J. opt. Soc. Am.* **42**, 286.

Lord, R. C. and McCubbin Jr., T. K. (1957). *J. opt. Soc. Am.* **47**, 689.

Low, F. J. (1961). *J. opt. Soc. Am.* **51**, 1300.

Martin, D. H. and Bloor, D. (1961). *Cryogenics* **1**, 159.

Martin, D. H., Bloor, D., Dean, T. J., Jones, G. O., Mawer, P. A., and Perry, C. H. (1961). *Proc. R. Soc.* **A260**, 510.

Michelson, A. A. (1902). "Light waves and their uses." University of Chicago Press, Chicago.

Möller, K. D., Tomaselli, V. P., Skube, L. R., and McKenna, B. K. (1965). *J. opt. Soc. Am.* **55**, 1233.

Oetjen, R. A., Haynie, W. H., Ward, W. M., Hansler, R. L., Schauwecker, H. E., and Bell, E. E. (1952). *J. opt. Soc. Am.* **42**, 559.
Papoular, R. (1964). *Infrared Phys.* **4**, 137.
Plyler, E. K., Yates, D. J. C., and Gebbie, H. A. (1962). *J. opt. Soc. Am.* **52**, 859.
Putley, E. H. (1965). *Appl. Optics* **4**, 649.
Renk, K. F. and Genzel, L. (1962). *Appl. Optics* **1**, 643.
Ressler, G. M. and Moller, K. D. (1967). *Appl. Phys.* **6**, 893.
Richards, P. L. (1964). *J. opt. Soc. Am.* **54**, 1474.
Richards, P. L. (1965). *Pure appl. Chem.* **11**, 535.
Shenker, H., Moore, W. J., and Swiggard, E. M. (1965). *J. appl. Phys.* **35**, 2965.
Smith, S. D. (1958). *J. opt. Soc. Am.* **48**, 43.
Strong, J. (1949). *J. opt. Soc. Am.* **39**, 320.
Tolansky, S. (1948). "Multiple-Beam Interferometry of Surfaces and Films." Oxford University Press, Oxford.
Turner, A. F., Chang, L., and Martin, T. P. (1965). *Appl. Optics* **4**, 927.
White, J. U. (1947). *J. opt. Soc. Am.* **37**, 713.
Yamada, Y., Mitsuishi, A., and Yoshinaga, H. (1962). *J. opt. Soc. Am.* **52**, 17.
Yoshinaga, H., Mitsuishi, A., Otsuka, Y., and Fujita, S. (1963). *J. appl. Phys. Japan* **2**, 574.
Weinstein, W. (1947). *J. opt. Soc. Am.* **37**, 576.

Bibliography

Martin, D. H. (ed.) (1967). "Spectroscopic Techniques for Far-Infrared, Sub-Millimetre and Millimetre Waves." North-Holland Publishing Company, Amsterdam.
Martin, A. E. (1966). "Infrared Instrumentation and Techniques." Elsevier Publishing Company, Amsterdam.
Smith, R. A., Jones, F. E., and Chasmar, R. P. (1968). "The Detection and Measurement of Infrared Radiation." 2nd Edition. Oxford University Press, Oxford.
Houghton, J. T. and Smith, S. D. (1966). "Infrared Physics." Oxford University Press, Oxford.

Chapter 3

TECHNIQUES

Most of the sampling techniques used in the far infrared are in principle no different from those employed in regions of higher frequency, and are well described in many texts (see, for example, Potts 1963). Differences result from the various transmission properties of materials and the reduced intensity of available radiation. In general, intensities are found to be lower than in the mid-infrared region and, hence, larger samples are often required.

I. Window Materials

The infrared transmission spectra of a number of materials are shown in Fig. 1. Before commenting on individual materials, it may be mentioned that transmission properties of optical crystals are strongly temperature dependent, and that the given spectra and subsequent remarks apply only to ambient temperatures. Hadni *et al.* (1965) have found that at liquid helium temperatures fifteen crystals, including several alkali metal thallium halides, which at room temperature were completely opaque, became transparent in the far infrared. Crystals

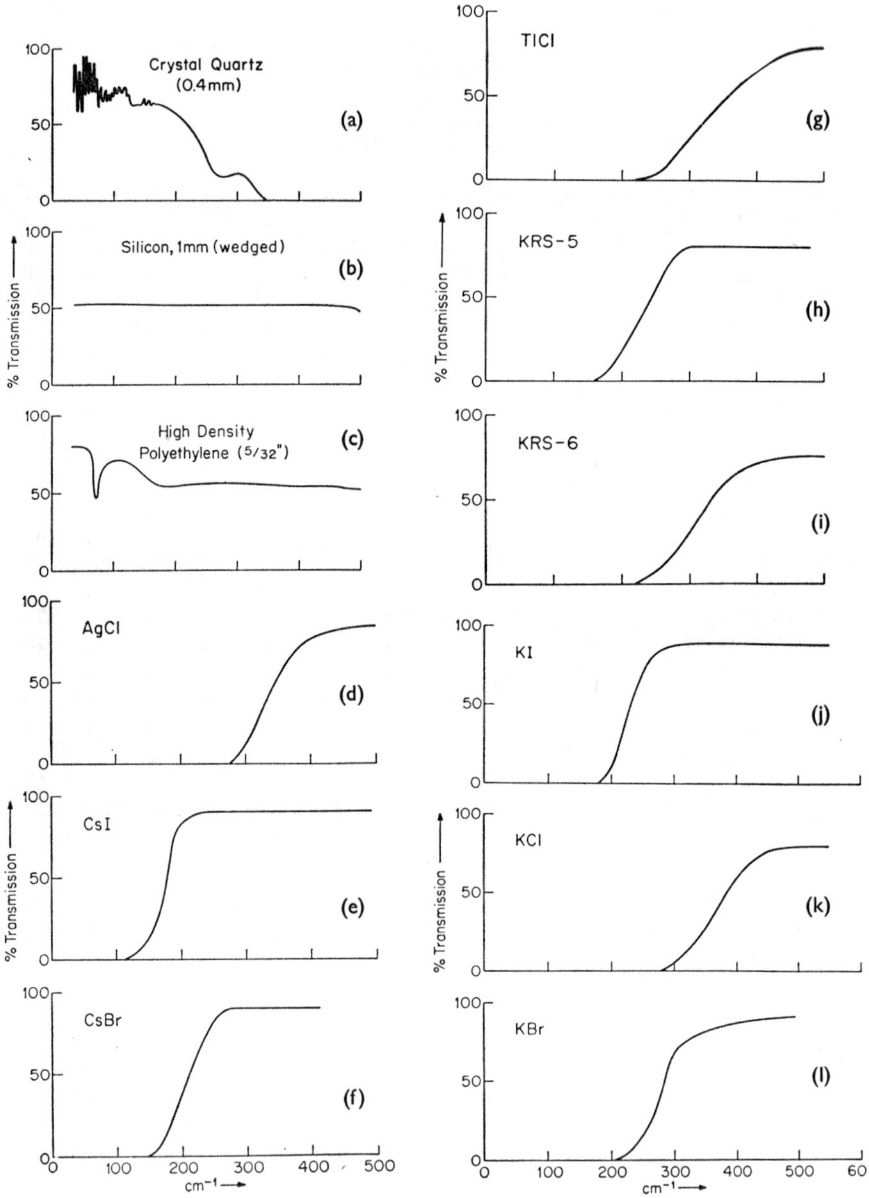

FIG. 1. Transmission of (a) crystal quartz cut so optical axis is parallel to the surface, (b) silicon cut with a 0° 50′ wedge, (c) high-density polyethylene $\frac{5}{32}$ inch, (d) AgCl, (e) CsI, (f) CsBr, (g) Kl, (h) KCl, (i) KBr, (j) TlCl, (k) KRS-5 and (l) KRS-6.

of about 3 mm thickness were employed, and in each case the onset of "supertransparency", sometimes to values of over 80 per cent transmission, was very sharp. It is therefore possible that these materials could be used as filters with "clean" cut-offs, or possibly be considered as window materials for particular cryostatic applications.

A. Polyethylene

This is undoubtedly the cheapest and most versatile window material in current use, and is commercially available in various high-density forms (for example Rigidex, Iridon, Marlex, and so on).

The transmission spectrum, given in Fig. 1c, shows a single sharp band at 72 cm^{-1}. However, it has been found by Stone and Williams (1966) that as polyethylene is cooled from room temperature to about 164°K, the absorption becomes more intense, and shifts to 78 cm^{-1}. Consequently, if polyethylene windows are used in a cell cooled to low temperatures in a double-beam spectrometer it becomes necessary to treat the reference cell similarly in order to compensate for this effect.

Where polyethylene is used in gas cells there is the possibility of adsorption and subsequent diffusion of the gas into the window. Since adsorbed molecules may show frequency shifts, relative to their position in the gas phase, the observed spectrum will be a superposition of the vapour and the condensed-state spectra. Studies on bromocyclobutane vapour by Rothschild (1965a, b) have shown that the frequency shifts are accompanied by considerable intensity changes and may result in pronounced apparent differences in the band contours, with the serious possibility of wrong assignments. Further, it is clearly possible to contaminate cell windows as a result of such adsorption, and unless testing procedures, such as a time-study of the spectrum, are applied, it is probably better to discard all polyethylene windows after use.

There are various mechanical disadvantages inherent in the use of polyethylene. In particular, if the spectrometer is evacuated the elasticity of the material is sufficient to cause outward "bowing" of the windows, with consequent uncertainty in the cell path-length. It is possible to use increased thicknesses of window material to counteract this, and plates up to $\frac{1}{4}$ inch thickness have been used by Willis *et al.* (1963); however, there is a parallel increase in the scattering of incident radiation and consequent reduction of energy reaching the detector.

B. Crystal Quartz

As can be seen from Fig. 1a, this material is suitable for the region below 200 cm^{-1}. It has the further advantage that it can act as a transmission filter between 250 and 2500 cm^{-1}. If an accurately known path length is required, crystal quartz is an excellent choice as it will not deform under stress. Slight wedging is sufficient to eliminate interference effects which can result from its high reflectivity.

C. Silicon

This material has been used relatively little to date but will probably increase in popularity despite rather high reflection losses. It is transparent over the entire far-infrared region (Fateley *et al.* 1966), unlike crystal quartz being completely free of interfering absorptions, and is considerably cheaper. It does not deform under stress like polyethylene although it has the disadvantage of being extremely brittle and breaking easily; it is both chemically inert and resistant to absorption of solvents or chemicals. Commercially obtainable wedged-cuts are available so that interference effects are not a problem. It is possible that it may be used in place of the extremely expensive diamond windows used in Golay detectors. Its use for reflectance spectra in the far infrared is being actively developed by Carlson *et al.* (1967), and this is discussed in Section V.

D. Other Materials

1. *Diamond*

Diamond windows are suitable for the entire 10–1000 cm^{-1} region and are used for specialized purposes, such as Golay windows, and as windows in special high-pressure cells.

2. *Inorganic Crystals*

Although opaque throughout much of the far-infrared region, certain crystals, for example sodium and lithium fluorides, show excellent transmission below 50 cm^{-1}, and their use here should be considered. A complete bibliography covering transmission properties of crystals has been compiled by Palik (1960).

3. *Irtrans*

Various polycrystalline compounds, designated Irtrans (numbers one to five), are supplied by the Eastman Kodak Company, and have been investigated as possible far-infrared filters or windows materials by McCarthy (1966) and Ressler and Möller (1966). Some relevant details are collected in Table I.

Irtran 5 especially would seem to be a promising material; its refractive index is less than that of Irtran 2, and hence losses by reflection are less, and McCarthy states that a 1 mm thick sample transmitted 70–80%

TABLE I. Properties and Compositions of Irtrans 1–5.

Irtran	Formula and structure	[a]Far infrared cut-in (cm^{-1})	[b]Per cent transmission at given wavenumber (cm^{-1}) 100	60	33
1	MgF_2: rutile	119	3	20	48
2	ZnS: 95% sphalerite 5% wurzite	111	4	25	40
3	CaF_2: fluorite	50	<1	2	23
4	ZnSe: blende	130	<1	<1	<1
5	MgO	100	5	46	55

[a] Taken from Ressler and Möller. [b] Taken from McCarthy. Crystals varied from 1–2 mm in thickness.

radiation from 50–33 cm^{-1}. Slight wedging suffices to reduce interference fringing.

II. Gas Cells

Gas cells are normally fitted with polyethylene windows. Any bowing of these under a pressure differential is not so important, as path lengths of 10 cm or greater are commonly used and so the relative error is small.

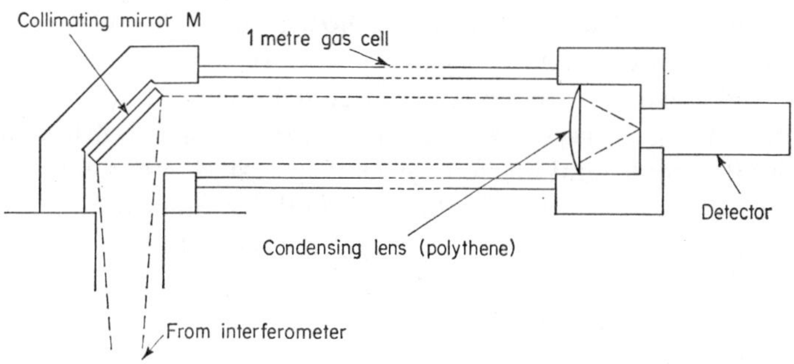

FIG. 2. Optical diagram for long-path gas cell for Grubb-Parsons NPL Cube interferometer.

The problem of long path cells is often the same as in the mid infrared, namely, that the cell must be tailored to fit the instrument, and in the past the available sample area was often such as to accept a maximum of some 10 cm. To achieve a greater path length may mean recourse to a multipass cell or extra optics to direct the radiation through the cell externally. The problem is particularly important in the far infrared not only because of the lack of source intensity but also because some modes of interest, for example torsions, are inherently weak.

Some interferometer systems are especially well suited to the use of long path cells. In principle, if the radiation emerges as a parallel beam from the interferometer, then a long cell can simply be interposed between the interferometer and the detector condensing lens. Figure 2 shows how this can be achieved in a Grubb Parsons—NPL Cube interferometer. The slightly divergent radiation emerging from the interferometer is rendered parallel by mirror *M* and passed down the gas cell where it is condensed on to the detector by the polyethylene lens.

III. Solids

Apart from single crystals, solids are usually handled as mulls in Nujol contained between polyethylene plates or as discs pressed from a mixture of the ground sample and powdered polyethylene. Since there is less concern over particle size than in the mid-infrared region, scattering losses being much lower, preparation of mulls is considerably easier.

If the sample is examined as a mull in an evacuated spectrometer, a convenient method is to use polyethylene windows with a spacer, depending on the path length required. If the windows deform, air bubbles may appear in the cell. This can be avoided by using thicker polyethylene. If the spacer is laid on one of the polyethylene plates and the window area filled with the mull, the top plate can be carefully laid on and air excluded. The cell can then be screwed down to obtain a vacuum tight seal in a convenient cell holder.

Various methods have been given for preparing polyethylene discs, involving different combinations of heat, pressure, and type and grade of powdered polyethylene. In one method by Smethurst and Steele (1964) a mixture of powdered polyethylene (about 150 mg) is ground up together with the sample (3–30 mg) and heated to 120°C while a pressure of about 6 tons in^{-1} is applied.

Sometimes the temperature may cause decomposition of the sample. If this occurs it has been found by May and Schwing (1963) that discs can be prepared by mixing paraffin wax (m.p. 55°C) with the polyethylene and pressing the discs at lower temperatures.

IV. Liquid Cells

Numerous techniques have been used for liquid cells, depending on the information required, the difficulty of handling liquid or solution samples, and whether or not the sample chamber is operated under vacuum or ambient pressure.

In general, the fabrication of suitable cells is simply obtained by adapting cells normally employed in the mid-infrared region to use the window materials mentioned earlier. Most commercially made spectrometers are freed from atmospheric water vapour by flushing with dry air or nitrogen; with such instruments, standard cells with polyethylene windows are commonly employed. However, many "homemade" grating spectrometers and most interferometers operate under vacuum so that sealing problems become acute. Use of, say, 1 or 2 mm thickness polyethylene cell windows is satisfactory only if band positions and relative intensities are required; the bowing mentioned previously precludes accurate measurements of absolute intensities or extinction coefficients. If these are necessary, then mechanically superior materials, for example KRS-5 above 200 cm^{-1} or crystal quartz below 200 cm^{-1}, are employed at the expense of the cheapness and extensive transparency of polyethylene.

V. Reflectance Spectra

For the infrared study of surface phenomena, for example adsorption, or for the examination of insoluble or otherwise intractable solids, for example some polymers, absorption methods are often difficult or impossible, and recourse is had to reflectance techniques. Since 1961, a particular technique of attenuated total reflectance (ATR) and also multiple attenuated total reflectance (MATR) introduced by Fahrenfort (1961, 1962) have been exploited (see Hansen, 1965, for discussion and further references). The principle involved is that the radiation, which is incident at an angle less than the critical angle, and directed at the interface between the sample and a medium which is both highly refractive and also transparent, is attenuated and emerges as a spectrum closely resembling a series of absorption bands. Excellent infrared spectra can be obtained, indices of refraction and absorption may be measured, and the technique is superior to conventional reflectance methods. By a development using multiple internal reflections (MATR) greater sensitivity can be obtained, and the spectra of very small amounts of sample measured.

The essential characteristics of a reflectance crystal are high transparency and high refractive index. Since low-frequency modes are often relatively weak, MATR is indicated. KRS-5 can be commercially obtained in suitable ATR crystal form and cell attachment (and has been successfully used down to 250 cm^{-1}).

The possible use of silicon as a MATR crystal has been investigated by Carlson (1967); it is transparent to more than 50% over the

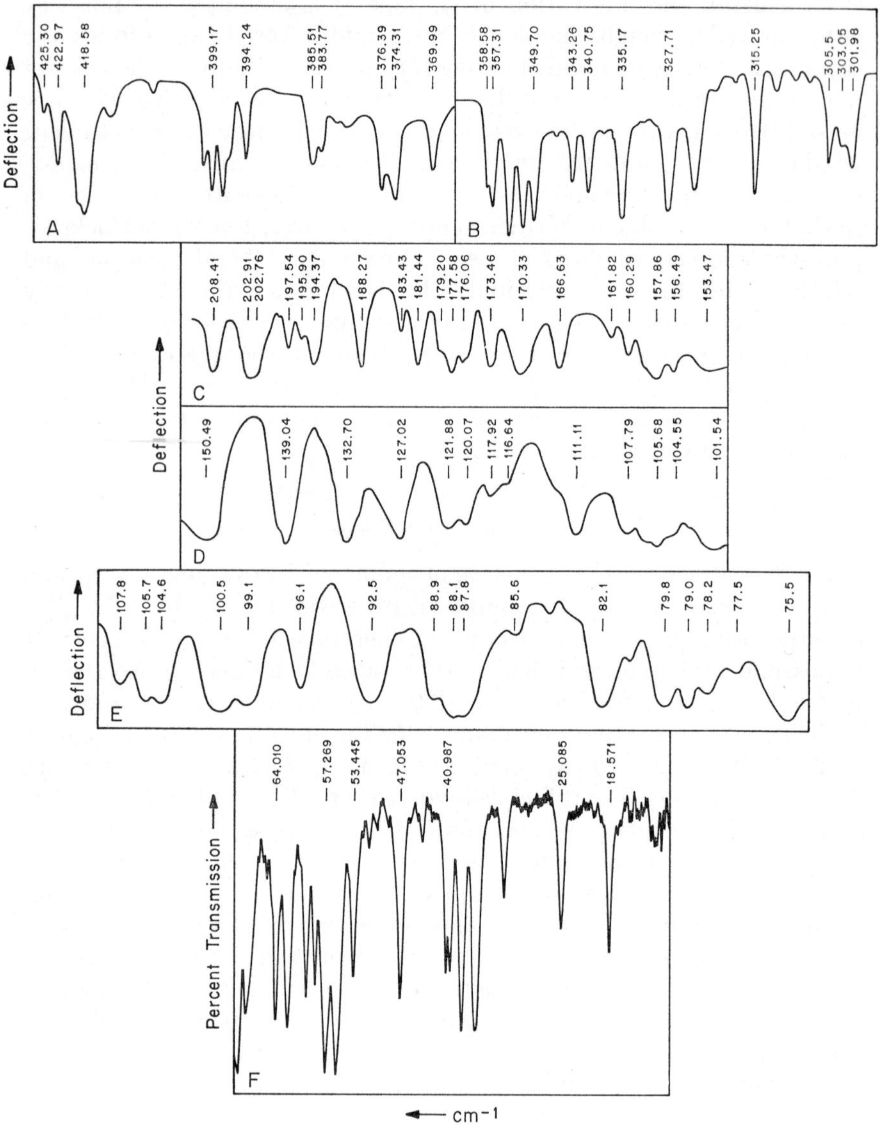

FIG. 3. Water vapour spectra from 425 to 10 cm^{-1}. A–E redrawn from Plyler and Acquista (1956); F redrawn from a spectrum supplied by Research and Industrial Instruments Co., London, England. A 320 lines per inch grating spectrometer; 3rd order spectrum. B 320 lines per inch grating spectrometer; 2nd order spectrum. The two orders overlap at 305 cm^{-1}. C 320 lines per inch grating spectrometer; 1st order spectrum. D 180 lines per inch grating spectrometer. E 180 lines per inch grating spectrometer. F Interferometric spectrum; resolution 0.5 cm^{-1}. Stated frequencies taken from Hall and Dowling (1967).

region below 250 cm^{-1} and has the high refractive index requirement, although reflection losses may be high. Several samples of crystalline silicon of various degrees of purity and with various dopants were tested for transmission, and then MATR measurements were attempted using a Wilks Model 9 attachment to a Beckman IR-11 grating spectrometer. More recently Carlson has tested a piece of specially cut silicon (Wilks Scientific Company) and found that MATR spectra could be obtained below 150 cm^{-1}. Preliminary results indicated that satisfactory MATR spectra of certain solid samples were possible down to 33 cm^{-1}.

VI. Calibration

A. Water

By far the most common procedure is to admit atmospheric water vapour to the sample chamber of the spectrometer or interferometer and use the pure rotation spectrum which extends over the range 10–400 cm^{-1}. Figure 3 illustrates various reported spectra for this region.

Rao *et al.* (1962, 1966) have discussed the precision and accuracy of this and other calibrants for the far-infrared region. The sixty pure rotation lines between 271 and 400 cm^{-1} derive from direct infrared measurements by Rao *et al.* (1962) and are known to an accuracy of ± 0.03 cm^{-1} for unblended lines. Lines between 120 and 271 cm^{-1} are calculated from energy level data, and given limits of error are ± 0.05 to ± 0.1 cm^{-1}. For greatest accuracy below 120 cm^{-1} they recommend use of various simple gases; these are listed in Section C below. However, for almost all chemical purposes the water vapour spectrum suffices.

B. Mercuric Oxide

If there is a large difference in resolution between the standard water vapour spectrum and the given spectrum, comparison is sometimes difficult and the process is tedious, involving measurements of many bands. McDevitt and Davidson (1965) have proposed the use of a simple secondary solid standard, analogous to the use of polystyrene at shorter wavelengths. This consists of yellow mercuric oxide (10% by weight) mixed with low-density polyethylene powder and pressed into disc form. Prepared under specified directions this results in a spectrum with five usable bands in the far- and mid-infrared region, as shown in Fig. 4. Band positions are given as 596 ± 1, 492 ± 1, 225 ± 2, 145 ± 0.5, and $60 \perp .5$ cm^{-1}; the standard is easily fabricated and can be stored.

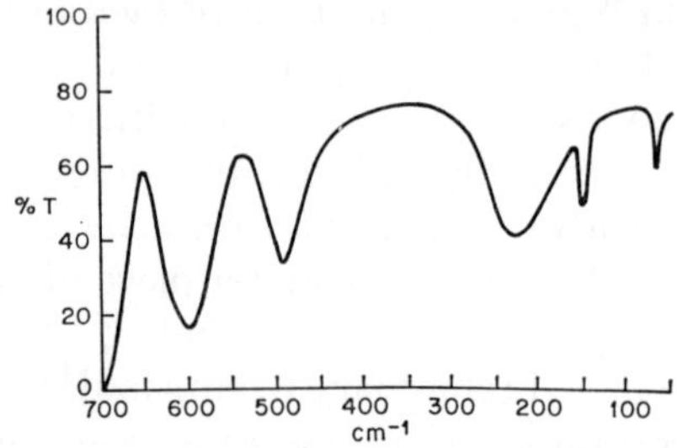

FIG. 4. Far-infrared spectrum of yellow mercuric oxide in polyethylene matrix.

C. Gases

For frequencies below 120 cm^{-1}, greatest accuracy is obtained using gases for which the ground state rotational constants are well established. "Calculated" pure rotation lines of carbon monoxide, hydrogen cyanide and nitrous oxide are available, and are listed in Table II.

TABLE II. Pure Rotational Lines of Carbon Dioxide, Nitrous Oxide, and Hydrogen Cyanide (cm^{-1}), from Rao *et al.* (1966).

J	$C^{12}O^{16}$	$N_2^{14}O^{16}$	$HC^{12}N^{14}$
0	3.845	0.838	2.956
1	7.690	1.676	5.913
2	11.534	2.514	8.869
3	15.379	3.352	11.825
4	19.222	4.190	14.781
5	23.065	5.028	17.736
6	26.907	5.866	20.691
7	30.748	6.704	23.646
8	34.588	7.542	26.599
9	38.426	8.380	29.553
10	42.263	9.217	32.505
11	46.098	10.055	35.457
12	49.932	10.893	38.408
13	53.763	11.730	41.358
14	57.593	12.568	44.307
15	61.420	13.405	47.255
16	65.245	14.243	50.202
17	69.068	15.080	53.148
18	72.888	15.918	56.092
19	76.705	16.755	59.036
20	80.519	17.592	61.977
21	84.330	18.429	64.918
22	88.138	19.266	67.856
23	91.943	20.103	70.793

TABLE II—*continued*

J	$C^{12}O^{16}$	$N_2^{14}O^{16}$	$HC^{12}N^{14}$
24	95.744	20.940	73.729
25	99.541	21.776	76.663
26	103.334	22.613	79.595
27	107.124	23.449	82.524
28	110.909	24.285	85.453
29	114.690	25.122	88.379
30	118.467	25.958	91.30
31		26.794	94.22
32		27.629	97.14
33		28.465	100.06
34		29.301	102.98
35		30.136	105.89
36		30.971	108.80
37		31.806	111.71
38		32.641	114.61
39		33.476	117.51
40		34.310	120.41
41		35.145	
42		35.979	
43		36.813	
44		37.647	
45		38.481	
46		39.314	
47		40.147	
48		40.980	
49		41.813	
50		42.646	

VII. Solvents

Depending on the length of the "window" required, there is considerable choice of solvents for use in the far-infrared region. Spectra published within the last few years include the following: carbon tetrachloride, chloroform, benzene and dioxane, from 5–500 cm^{-1} (Stanevich and Yaroslavski 1961); some substituted benzene derivatives from 20–420 cm^{-1} (Green *et. al.*, 1963); twelve liquids, including chloro- and bromobenzene, methylene chloride, carbon disulphide, *n*-hexane, cyclohexane, toluene and Nujol, from 50–650 cm^{-1} (Wyss *et al.* 1964). Figure 5 shows some spectra taken from the latter paper; measurements were made using caesium iodide windows to 170 cm^{-1} and Teflon or quartz for the lower frequency spectra. These authors stated that polyethylene was unsatisfactory, the windows being strongly

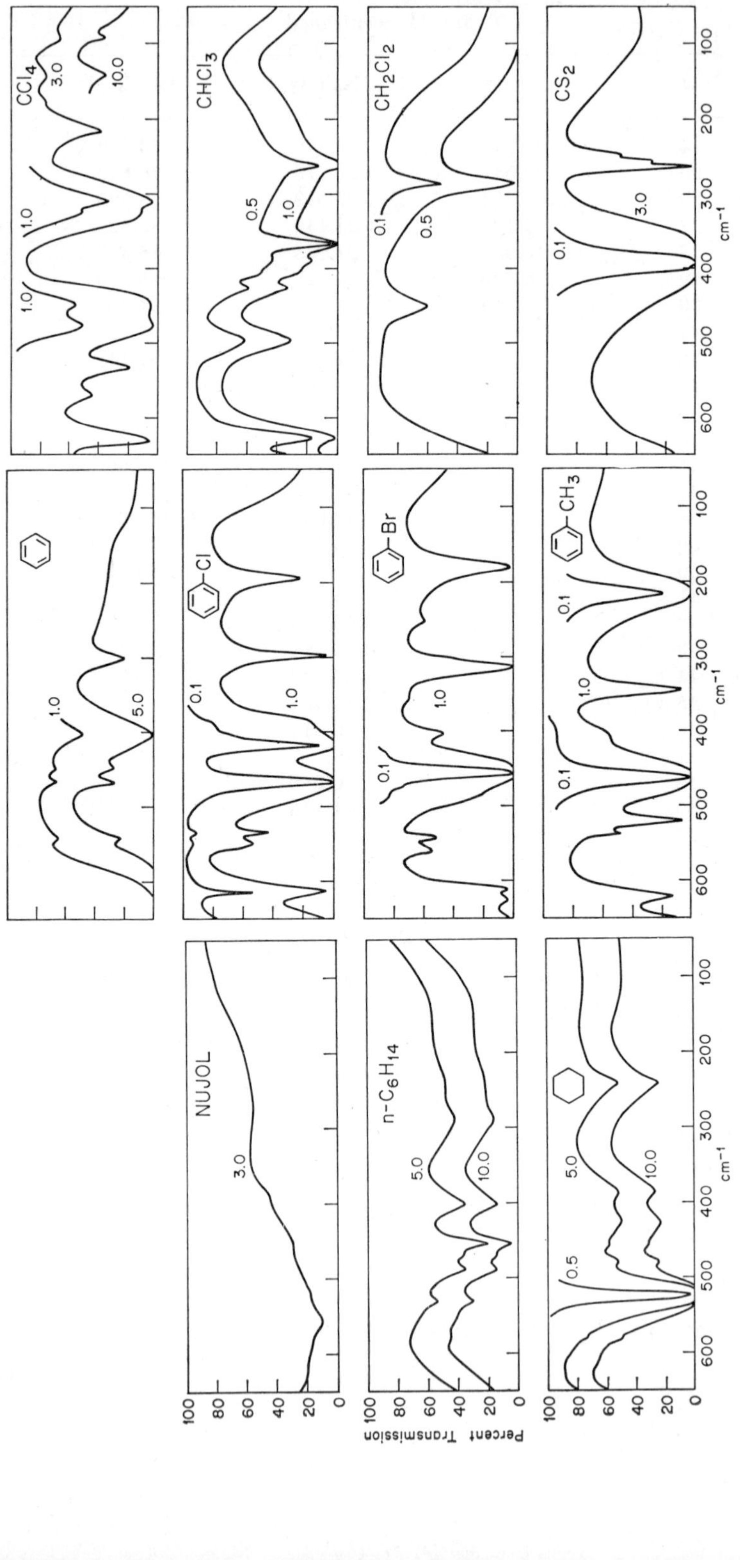

FIG. 5. Transmission of various liquids used as solvents in the far infrared. Numbers refer to thickness in mm.

swelled by most solvents, but in a later note they reported that high-density polyethylene (Rigidex Type 35, from British Resin Products Ltd) was excellent for the purpose.

A note on the difference between gas- and liquid-phase spectra is given in Section XI.

VIII. Matrix Isolation

This technique, of which polyethylene dilution (*vide infra*) is a rather specialized example, is, like most others mentioned in this chapter, not specifically devised for the low-frequency region. However, since its use there will probably increase, and since it receives little note in many texts, a brief mention will be made. The method is designed to permit spectral studies of compounds which, under normal conditions, are too unstable for isolation or examination. A typical recent example from Noble and Pimentel (1968), to whom much of the technique is due, is the first synthesis and vibrational analysis of hypofluorous acid, HOF, in a nitrogen matrix at 14° and 20°K. The object is to isolate reactive molecules in a solid inert environment at a temperature sufficiently low to prevent effective translational movement. The efficiency of the matrix in a monomer/polymer system, such as a hydrogen-bonded species, has been defined as the ratio of the numbers of molecules in the monomer to associated states. Noble has suggested that the ratio of the peak absorbance of the band assigned to the monomer to the sum of the absorbances of bands due to associated molecules represents the lower limit of the fraction of monomer molecules. The problem of frequency phase changes is difficult to evaluate. However, assuming the relation

$$\frac{\nu_0 - \nu}{\nu_0} = C\frac{D-1}{2D+1}$$

between the frequencies in the gas phase and condensed phase, ν_0, ν, respectively, and the dielectric constant of the medium, D, then shifts for argon and nitrogen matrices are of the order of only 1 per cent.

A. Properties of Matrices

1. *Inertness*

Obviously the medium must be chemically inert, and the rare gases are ideal. Hydrogen and nitrogen are also often suitable and, depending on the sample, many others may be considered.

2. *Optical Transmission*

In addition to the necessity of transparency in the spectral region required, undue scattering must be avoided. Thus, glassy media are preferred to crystalline, and temperature control may be important. Xenon, for example, is transparent at 66°K and scatters radiation below 50°K. In general, scattering is reduced by using very slow rates of deposition, although this is not always feasible in practice.

3. *Volatility*

The matrix must be both sufficiently volatile to allow manipulation *in vacuo* at ambient temperatures and also have a vapour pressure low enough at the refrigerant temperature to prevent sublimation. This requirement sets upper limits to the temperature ranges at which operation is possible; for example 70°, 35° and 30°K for xenon, argon and nitrogen, respectively.

4. *Rigidity*

By this is meant the ability to resist diffusion of the entrained sample. Diffusion is highly temperature-dependent (sometimes exponentially) and very low temperatures are often needed (about 20°K); this is one of the main experimental drawbacks to use of the technique.

B. Sample Preparation and Adsorption

There are two types of processes, corresponding to forming the active species and trapping immediately before subsequent decomposition, and generating the active species by suitable excitation, in practice almost always photolytically within the matrix. The former method may involve pyrolysis, dissociation in a glow- or arc-discharge, or suitable chemical reaction, and is accomplished by spraying a gas mixture of both matrix and active species on to a cooled surface. If the mole ratio of matrix to active species is sufficiently high, the matrix will isolate trapped molecules and prevent recombination reactions. This trapping may even occur when the trapped material would not condense at the temperature of the matrix deposition, for example methane (m.p. 90°K) may be trapped by a carbon tetrachloride matrix maintained at temperatures considerably in excess of the freezing point (90°K).

IX. High-pressure Studies

Application of high pressures to vibrational studies generally (by Drickamer *et al.* 1957) and in low-frequency regions in particular (by

Lippincott *et al.* 1959, 1960, 1961) have been made for many years. Very recently, however, there has been increased activity, especially in the far-infrared region; and the current position is well summarized by Ferraro (1968), who has considered pressures up to 160,000 atmospheres.

There are several reasons for high-pressure studies, but probably the most important is as a means of distinguishing between intramolecular and intermolecular modes, for example hydrogen-bonding, lattice modes, and so on. At pressures of the order of 40,000 atmospheres, the decrease in volume of organic materials can amount to some 20 per cent; the space group symmetry of the unit cell may well change, and the intermolecular forces be perturbed with appropriate changes in relevant absorptions. There is relatively little change in the positions of most normal intramolecular absorption bands, although changes in intensity do occur in this pressure region; and Ferraro cites the use of this latter effect to distinguish between antisymmetric and symmetric stretching vibrations (for example metal–halogen modes) in polycrystalline materials (Postmus *et al.* 1967). Lattice modes of ionic solids shift to higher frequencies with increasing pressure, the magnitude depending on the compressibility of the crystal, in accordance with the formula:

$$\gamma\chi\nu = \left(\frac{\partial\nu}{\partial\rho}\right)_T$$

where χ is the isothermal compressibility, ν is the frequency, and γ is the Grüneisen parameter, which is essentially constant over a wide range of conditions and is itself given by the relation: $\gamma = V(\partial U/\partial P)_V$; typical experimental values (Slater 1940) of γ are: KF, 1.45; KCl, 1.60; KBr, 1.68; KI, 1.63. Hence, frequency shifts may be too small to be recorded for lattice modes of particular solids.

New developments of this technique have recently been outlined by Brasch and Jacobsen (1965a) and Brasch (1965) using a diamond cell to over 30 kbar. Initial application of pressure to organic liquids, for example benzene, results in a mass of crystallites, which usually exhibit polymorphism, and hence complicate any interpretation of spectra. Gradual release of pressure until all but one crystallite had melted, followed by increase of pressure, led to the production of a single crystal with controllable growth. All operations were followed microscopically, and the entire operation took less than one hour. Single crystals of rotational isomers of halogenated ethanes were studied spectroscopically in this way, and it was shown that the crystals were not always the same as those formed by the normal method of freezing at ambient pressure. Thus, for 1,1,2,2-tetrachloroethane, low-temperature solidification always produces the gauche form; pressure-formed crystals, however, are always *trans*, as predicted by normal coordinate calculations. A similar situation was noted for the corresponding bromo compound, but apparently in one instance the pressure method did lead to a single crystal of the gauche rotamer. With gauche crystals of both compounds, solid-phase transitions to the *trans* isomers were achieved by warming or by pressure reduction. Thus, the method seems well adapted to the study of solid-phase transitions. "Superpressurizing", analogous to supercooling, was observed, as with tetrachloroethane pressures of the order of 32 kbar were required for solidification at ambient temperature, while melting did not occur until the pressure had been reduced to about 15 kbar.

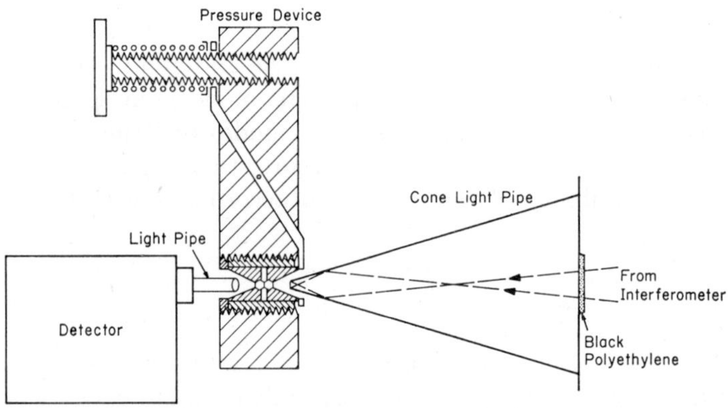

FIG. 6. Optical diagram of high-pressure system coupled to an interferometer.

A critical discussion of the various methods available for measuring the spectra of solids, not necessarily in the far-infrared region, has been given by Lippincott *et al.* (1961 *vide supra*). Advantages with the diamond cell include the small quantity requirements of the cell, the chemical inertness of diamond, and relative rapidity of operation. However, there are instrumental difficulties with the beam condensation requirements imposed by the small surface area of the diamonds. Two microcells employed in the Lippincott model had surface areas of 1.56 and 1.82×10^{-4} in^2, respectively; and the large cell used by Ferraro and co-workers was about 3.10^{-2} in^2.

Two groups (Fateley *et al.* 1967 and Gebbie *et al.* 1966) have discussed the techniques and instrumentation required to adapt high-pressure cells to interferometric measurement. The latter group employed two types of high-pressure apparatus. In one, a six-anvil cube type adapted from a design described by Houck and Johnson (1962) used crystal quartz windows, transmitting from 200–10 cm^{-1} (except for the adsorption near 125 cm^{-1}) and operated to 35 kbar pressure. In the other, a hydrostatic piston and cylinder device, similar to the Drickamer system, was usable to 12 kbar, and the relative merits of each apparatus are discussed.

The former group studied both inorganic and organic compounds in a cell with diamond windows from 360–40 cm^{-1} with pressures usually less than 50 kbar, and the results were compared with those obtained at ambient pressures and low temperatures. The cell and pressure device were supplied by High Pressure Optics, Inc. (McLean, Virginia, U.S.A.). The optical system is shown diagrammatically in Fig. 6. Emergent radiation from an FS-520 Fourier Spectrophotometer (Research and Industrial Instruments Co.) traverses a black polyethylene window into a brass cone light pipe of angle 9° 37′, and small end opening sixty thousandths of an inch, which is machined to fit the diamond cell piston. Radiation finally travels to the detector *via* the attached light pipe. The cell can be mechanically adjusted to an optimum position with respect to the light path. Water vapour was removed from the cone by flushing with dry air. The optics were checked by running spectra of known compounds (anthracene and ytterbium aluminum garnet) at atmospheric pressure.

Some of the results are of interest in discussing this technique and are shown in Table III and Fig. 7, which show the change in frequency of the transverse optical mode of various alkali halides.

TABLE III. Change in Frequency $\Delta\bar{\nu}$ as a Function of Pressure for Alkali Halides.

Crystal	1 atmosphere (cm^{-1})	42 kbar (cm^{-1})	$\Delta\bar{\nu}$
NaF	250	304	54
NaCl	165	218	53
KCl	145	162	33
KBr	122	140	18
KI	103	106	3

Table IV shows the different effects upon vibrational frequency by lattice compression with temperature as compared with pressure stress for crystals of zirconia and hafnia.

It will be noticed that the temperature and pressure effects sometimes differ in direction, cooling always producing an increase in frequency. Thus, as noted by Brasch (*vide supra*) for organic compounds, no consistent pattern emerges at the present time.

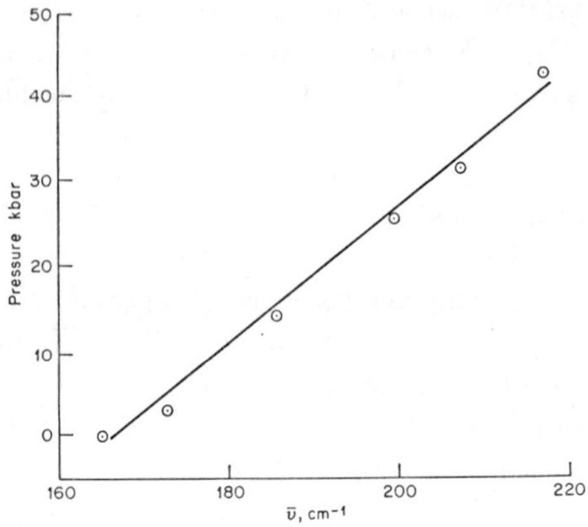

FIG. 7. Plot of frequency *versus* pressure for the transverse optical mode of sodium chloride.

In a discussion of pressure calibration techniques, the important point is made that there is an apparently unavoidable pressure gradient across the pressure face of the diamonds.

TABLE IV. Pressure and Temperature Effect upon Frequency for ZrO_2 and HfO_2.

Crystal	1 atmosphere, 25° (cm^{-1})	40 kbar, 25° (cm^{-1})	$\Delta\bar{\nu}$ (25°)	$\Delta\bar{\nu}$ (−123°)
ZrO_2	272	270	−2	+3
	262	259	−3	+2
	233	235	+2	+8
HfO_2	270	270	0	+3
	260	258	−2	+1
	237	242	+5	+7

X. Polyethylene Dilution

This is an elegant technique, due to Brasch and Jacobsen (1964, 1965) which is quite generally applicable to other regions of the spectrum but is particularly useful in the study of weak interactions, such as hydrogen bonds, in the far-infrared range. Incorporation of samples into pressed discs of suitable alkali halides, for example caesium iodide, is well known; this method involves rather similar containment within a polyethylene matrix. Commercial *low*-density polyethylene pellets (for example Alathon 10, available from du Pont de Nemours and Company) were formed into discs by application of heat (150°) and pressure (30,000 lb thrust for 10 min). Liquid samples boiling over 100° were simply incorporated by covering a disc and heating to about 110°; at this temperature the translucent polyethylene becomes transparent and absorbs easily. The amount retained is controlled by the time of soaking and heating; a few seconds is stated to be sufficient for normal spectroscopic work, and excess can be expelled by reheating. Low boiling samples are treated by using an autoclave, and solids are melted and the polyethylene dipped.

Spectra obtained using this technique for several different types of compounds run at ambient or at liquid-nitrogen temperatures each correspond to "normal" solution spectra at ambient temperatures. There is thus the possibility of avoiding the complications of lattice modes, which might arise in low-temperature spectra studies of the pure sample, as well as solvent effects with conventional solvents.

This has been exploited in a study of hydrogen bonding in phenols (*vide supra*, Brasch and Jacobsen 1965) in the range 4000–70 cm^{-1}. The principle of the method was to follow the disappearance of relevant modes by recording spectra of successively diluted phenol solutions. Such absorptions are inherently weak, and difficulties of measurement are aggravated in the far infrared; the extreme dilutions needed require very long path lengths and use of common solvents poses enormous experimental problems. The adsorbed phenol was shown to be present initially as a hydrogen-bonded polymer, and gentle heating gradually converted this into the monomeric species. Cooling the monomeric form to −190° restored the polymeric state, that is "dilution" studies are possible by varying temperature rather than concentration. In addition, the –OH torsional mode round the –CO bond in the monomer was located near 300 cm^{-1}.

XI. Gas-liquid Frequency Shifts

Although this is not a point of technique, whether or not a given sample is measured in the liquid- or gas-phase often depends on the instrumentation and techniques available, and hence it will be considered here. It is well known that in the higher-frequency infrared region, 700–4000 cm^{-1}, gas-phase frequencies are usually higher, by some 10 cm^{-1} or more, than the corresponding liquid-phase frequencies (Kagarise 1963). It appears that below about 300 cm^{-1}, however, the reverse situation may obtain, and the gas frequencies are the smaller (Fateley *et al.* 1964). These authors have shown that in a study on twenty-six compounds, all except three had $\nu_g < \nu_{liq}$ and the exceptions were in the region above 350 cm^{-1}. Some examples are given below in

Table V. Frequencies were obtained for each phase using the same instrument, and the major uncertainty is stated to be the selection of the band centre of some broad bands.

TABLE V. Comparison of Gas-phase and Liquid-phase Frequencies.

Compound	ν_g (cm^{-1})	ν_{liq}	$\nu_g - \nu_{liq}$
Toluene	341.0	346.2	−5.2
	205.1	215.7	−10.6
o-Xylene	247 ± 1	258 ± 2	−11 ± 3
p-Xylene	132.0	149 ± 1	−17 ± 1
Benzaldehyde	111.0	133 ± 2	−22 ± 2
3-Methylpyridine	336.5	340.8	−4.3
	194.2	206 ± 2	−12 ± 2
Dimethyl sulphide	172.3	175 ± 3	−3 ± 3

The authors of the above study make the interesting point that since the shift in frequencies appears to change sign, depending on the spectral region, there should be a point, or range, where $\nu_g = \nu_{liq}$; this appears to be in the range about 350–750 cm^{-1}.

Perhaps the most important consequence is the effect of such changes on calculated thermodynamic functions. As stressed in Chapter 1, low frequencies have much greater weight than those in higher ranges of the spectrum. In general, thermodynamic values calculated from liquid-phase low-frequency assignments will be too low, which is the reverse of what has usually been assumed by analogy from changes in higher-frequency regions. The point is important, because low-frequency data are commonly taken from Raman measurements which still only rarely derive from gas-phase studies, and it is possible that a similar effect may occur. For the special example of the calculation of the barrier to internal rotation, the error may become enormous since, at least in the harmonic oscillator calculation, the frequency term is squared. Lockhart (1966) has estimated that a 20 per cent change in assigned vibrational assignments could affect the height of the rotation barrier in phenyl boron dichloride by several kilocalories. Presumably similar results apply for solution spectra, although there are few results available for comparison.

References

Brasch, J. W. (1965). *J. Chem. Phys.* **43**, 3473.

Brasch, J. W. and Jakobsen, R. J. (1964). *Spectrochim. Acta.* **20**, 1644.

Brasch, J. W., (1965). *Spectrochim. Acta.* **21**, 1183.

Brasch, J. W. and Jakobsen, R. J. (1965). *Spectrochim. Acta.* **21**, 1753.

Carlson, G. L., (1967). "Investigation of the Feasibility of Attenuated Total Reflectance Measurements in the Far Infrared Region" (McGovern, J. J. and Fateley, W. G., Eds.) AFML TR-67-303 p. 12.

Drickamer, H. G., Fitch, R. A., and Slyhuis, T. E. (1957). *J. Opt. soc. Am.* **47**, 1015.
Fahrenfort, J. (1961). *Spectrochim. Acta.* **17**, 698.
Fahrenfort, J. and Visser, W. M. (1962). *Spectrochim. Acta.* **18**, 1103.
Fateley, W. G., Matsubara, I., and Witkowski, R. E. (1964). *Spectrochim. Acta.* **20**, 1461.
Fateley, W. G., Witkowski, R. E., and Carlson, G. L. (1966). *Appl. Spectrosc.* **20**, 190.
Fateley, W. G., Witkowski, R. E., and McDevitt, N. T. (1967). Abstracts, 13th Colloqium, Spectroscopium Internationale, June 1967. Canada.
Ferraro, J. R. (1968). *Analyt. Chem.* **40**, 24A, and references therein.
Gebbie, H. A., Bradley, C. C., Gilbey, A. C., Kechin, E. V., and King, J. H. (1966). *Nature, Lond.* **211**, 839.
Green, J. H. S., Kynaston, W., and Gebbie, H. A. (1963). *Spectrochim. Acta.* **19**, 807.
Hadni, A., Claudel, J., Gerbaux, X., Morlot, G., and Munier, J-M. (1965). *Appl. Optics.* **4**, 487.
Hall, R. T. and Dowling, J. M. (1967). *J. Chem. Phys.* **47**, 2454.
Hansen, W. N. (1965). *Spectrochim. Acta.* **21**, 815.
Houck, J. C. and Johnson, U. O. (1962). *In* "High Pressure Measurement". (Giardini, A. A. and Lloyd, E. C. eds.) Butterworths, London.
Kagarise, R. E. (1963). *Spectrochim. Acta.* **19**, 1979.
Lippincott, E. R., Weir, C. E., Van Valkenburg, A., and Bunting, E. N. (1959). *J. Res. natn. Bur. Stand.* **63A**, 55.
Lippincott, E. R., Weir, C. E., Van Valkenburg, A., and Bunting, E. N. (1960). *Spectrochim. Acta.* **16**, 58.
Lippincott, E. R., Welsh, F. E., and Weir, C. E. (1961). *Analyt. Chem.* **33**, 137.
Lockhart, J. C. (1966). *J. Chem. Soc.* (A), 1552.
May, L. and Schwing, K. J. (1963). *Appl. Spectrosc.* 17, 166.
McCarthy, D. E. (1966). *Appl. Optics* **5**, 472.
McDevitt, N. T. and Davidson, A. D. (1965). *J. opt. Soc. Am.* **55**, 1695.
Noble, P. N. and Pimentel, G. C. (1968). *Spectrochim. Acta.* **24**, 797.
Palik, E. D. (1960). "A Far Infrared Bibliography". Natl. Res. Lab. Bibliog. No. 21, U.S. Dept. Commerce, Office of Tech. Service.
Plyler, E. K. and Acquista, N. (1956). *J. Res. natn. Bur. Stand.* **56**, 149.
Postmus, C., Nakamoto, K., and Ferraro, J. R. (1967). *Inorg. Chem.* **6**, 2194.
Potts, W. J. (1963). "Chemical Infrared Spectroscopy". John Wiley, New York.
Rao, K. N., Brim, W. W., Sinnett, V. L., and Wilson, R. H. (1962). *J. opt. Soc. Am.* **52**, 862.
Rao, K. N., Humphreys, C. J., and Rank, D. H. (1966). "Wavelength Standards in the Infrared". Academic Press, New York.
Ressler, G. M. and Möller, K. D. (1966). *Appl. Optics* **5**, 877.
Rothschild, W. G. (1965a). *J. Chem. Phys.* **42**, 694.
Rothschild, W. G. (1965b). *Spectrochim. Acta.* **21**, 852.
Slater, J. C. (1940). *Phys. Rev.* **57**, 744.
Smethurst, B. F. and Steele, D. (1964). *Spectrochim. Acta.* **20**, 242.
Stanevich, A. E. and Yaroslavski, N. G. (1961). *Optika Spektrosk.* **11**, 61.
Stone, N. W. B. and Williams, D. (1966). *Appl. Optics* **5**, 353.
Willis, H. A., Miller, R. G. J., Adams, D. M., and Gebbie, H. A. (1963). *Spectrochim. Acta.* **19**, 1457.
Wyss, H. R., Werder, R. D., and Gunthard, H. S. (1964). *Spectrochim. Acta.* **20**, 573.

Chapter 4

BARRIERS TO INTERNAL ROTATION: TORSIONS

In the 1930's, thermodynamic data provided conclusive evidence that in most molecules potential barriers had to be surmounted when turning from one configuration to another. Prior to this time the failure to isolate isomers arising from different orientations about a C–C bond had led to the view that rotation about a single bond was essentially free. In general there is a dependence of the potential, V, on the orientation of the parts of a molecule which can rotate relative to one another. The minima in the curve correspond to the more stable configurations of the molecule. The differences in potential energy between neighbouring maxima and minima are known as barrier heights, while the potential energy differences between the various minima are termed energy differences between isomers.

In general three types of behaviour may be recognized.

(1) If all barrier heights are large with respect to kT, then a molecule with a configuration corresponding to one minimum will be unable to convert, at a measurable rate, into another stable configuration. Different isomers then exist (one for each minimum) which may be separated from one another. Restricted rotation about double bonds is commonly of this type, giving rise to *cis* and *trans* isomers.

(2) If one or more of the barrier heights is comparable with kT it will

be possible for one stable configuration to convert rapidly into another. A dynamic equilibrium of rotational isomers (one for each minimum) is established. Restricted rotation about a single bond is commonly of this type.

(3) If changes in the potential energy are small with respect to kT then essentially free rotation occurs, for example the methyl groups in dimethylacetylene.

The problem of determining the potential energy barriers about single bonds has been the subject of investigation for several years. The far infrared, particularly the region below 200 cm^{-1}, is amenable to the study of barriers about single bonds, in that they are usually less than 5 kcal mole^{-1}, and the associated torsional vibrations arising from the internal barrier can often be observed in this region. Many contributions to this field have been made by microwave spectroscopy although this technique is somewhat limited by the high-frequency limit of about 10 cm^{-1}. The microwave and infrared methods may be viewed as complementary, however, in that the former method is best suited to molecules in which the torsion axis joins light groups while infrared techniques are best employed when heavy groups are involved.

I. Assignment of Torsional Modes

The overall problem of the determination of barriers to internal rotation in the far infrared rests with the accurate assignment of the torsional frequency. Theoretically, if the barrier is sufficiently high, there will be several energy levels corresponding to various torsional states in the molecule. It is the transitions from the ground state to the first or second of these levels which are observed and from which the height of the barrier can be computed (if its shape is known or is assumed to be known). Accurate assignment of the torsional frequency is often difficult. Although for many molecules in the vapour state the torsion may be the only absorption below 200 cm^{-1}, bending and other molecular modes, as well as difference tones, may appear in this region, especially in molecules containing heavy atoms. For the latter case, recourse to other means must be made for assignment. Mass changes or positional changes (other than in the group under study) will affect the position of low-frequency bending modes but may have little effect on the torsion. In the case of halogen substituted benzaldehydes, for example, Miller *et al.* (1967) have observed the torsion to consistently fall in the region 111–70 cm^{-1}, while the bending mode, found above 150 cm^{-1}, could easily be identified by its frequency dependence on both mass and position of the substituted halogen group.

It is often possible on the basis of deuterium substitution to assign a suspect low-frequency band. For example, if an absorption at 150 cm^{-1} was in question, a relatively simple calculation (which will be discussed in a later section) may be made to approximately determine the barrier height (providing the geometry of the molecule is reasonably well known—and without such information barrier studies will not proceed very far). Since the barrier height is little affected by deuterium substitution, a reverse calculation for the deuterium substituted molecule may be performed, yielding a new torsional frequency. Inspection of the substituted and unsubstituted spectrum with regard to frequency shift of the suspect band, should clearly indicate whether the absorption is the torsion.

The answer to the question of torsional assignments, however, is not as precise as may be desired. The most recent literature on the subject seems to indicate that the assignments are treated on the basis of individual molecules or classes of molecules. Microwave data is, of course, of considerable benefit; however, all molecules of interest have not as yet been studied in that many do not possess a dipole and, therefore, this information is somewhat limited.

Mention should be made of the assignment of torsions on the basis of "hot bands". These are absorptions which appear as a result of transitions between higher energy levels (see Fig. 4), that is $v_1 \rightarrow v_2$, $v_2 \rightarrow v_3$, and so on. Although they are not commonly observed, they can be of considerable aid in establishing unequivocally the torsional frequency. The calculated barrier, obviously, is the same for each transition and this test applied to a series of "hot bands" will establish the validity of the torsional assignment. Fateley and Miller (1963) succeeded in observing transitions between higher torsional levels for ethyl chloride and several other compounds. Calculations based on those observations provided an excellent test of the cosine potential energy function.

II. Factors Influencing Torsional Frequencies

Since barrier height calculations are based on accurate torsional frequencies, both the physical and chemical factors which influence these frequencies are of importance. The physical state of the sample has been shown by Fateley *et al.* (1964) and also by Radcliffe and Wood (1966) to have marked influence on far-infrared frequencies below 300 cm^{-1}. In general, gas phase frequencies have been observed to be 20–25% less than those in the liquid. In the case of the torsions, the frequency is squared in the actual potential barrier calculation and, consequently, serious errors are introduced if the correct frequency is not used. It is believed that barriers calculated from gas phase frequencies are inherently more accurate than those measured in the liquid in that the latter case may introduce proximity effects due to neighbouring molecules, for example van der Waals interactions, hydrogen bonding, and other environmental effects. Table I gives a series of torsional frequencies measured in both vapour and liquid states and the barriers calculated from these frequencies. It can be seen that relatively large errors in barrier heights can result from relatively small changes in torsional frequencies.

TABLE I. Torsional Frequencies and Barrier Heights for Benzaldehyde and Para-Substituted Benzaldehydes.

	Vapour ν torsion (cm^{-1})	V_2 (kcal $mole^{-1}$)	Liquid ν torsion (cm^{-1})	V_2 (kcal $mole^{-1}$)
Benzaldehyde	111.0	4.66	133	6.68
p-F-benzaldehyde	93.5	3.58	113	5.23
p-Cl-benzaldehyde	81.5	2.81	104	4.58
p-Br-benzaldehyde	73.5	2.37	92 ± 2	3.72
p-CH_3-benzaldehyde	89.5	3.47	114 ± 4	5.63

Intramolecular effects will also influence the barrier height and, consequently, the torsional frequency. The electronic character of the bond between the two groups in question will in part determine the energy of the barrier. For example, Table I gives both the torsional frequency and the barrier height for *para*-halogen substituted benzaldehydes and clearly shows a systematic change with substituent. Since steric effects are of no consequence in the *para* position, the observed changes are clearly the result of π interaction between the ring and the aldehyde group. Other obvious intramolecular effects, including hydrogen bonding and steric interactions, will influence the barrier height significantly and in reality it is the sum of all these interactions which is primarily responsible for the creation of the barrier.

III. General Theory

The calculation of the potential barrier from a torsional frequency requires certain specific information (and assumptions) relating to the structure of the molecule. Solution of the wave equation for the torsional energy of a molecule assumes that the potential energy may be adequately described by a Fourier series in terms of α, the angle of internal rotation, where

$$2V = \sum_{n=1}^{n=\infty} V_n(1 - \cos n\alpha) \tag{1}$$

or

$$V(\alpha) = \frac{V_1}{2}(1 - \cos \alpha) + \frac{V_2}{2}(1 - \cos 2\alpha) + \frac{V_3}{2}(1 - \cos 3\alpha)\ldots$$
$$+ \frac{V_n}{2}(1 - \cos n\alpha). \tag{2}$$

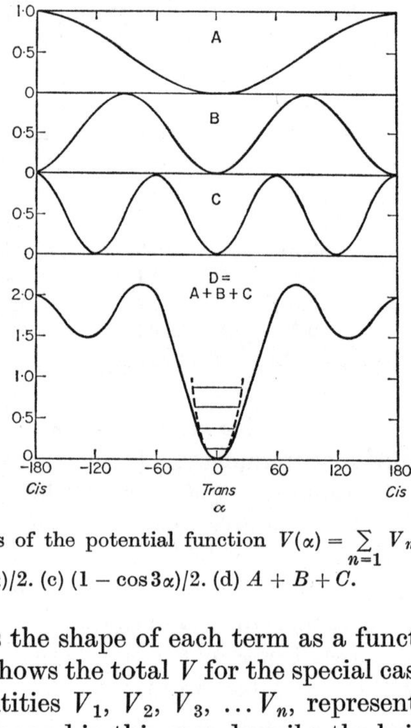

FIG. 1. Components of the potential function $V(\alpha) = \sum_{n=1} V_n(1 - \cos n\alpha)/2$. (a) $(1 - \cos \alpha)/2$. (b) $(1 - \cos 2\alpha)/2$. (c) $(1 - \cos 3\alpha)/2$. (d) $A + B + C$.

Figure 1 describes the shape of each term as a function of the internal angle α, and also shows the total V for the special case where $V_1 = V_2 = V_3 = 1$. The quantities V_1, V_2, V_3, ... V_n, represent the amplitude of the cosine functions and in this case describe the height of the barriers. The decision must be made as to which form of the potential functions is best suited for the molecule in question.

In the case of a rigid symmetric top (that is, a methyl group) bonded to a rigid framework which may be asymmetrical, the top has a three-fold symmetry axis. The potential energy of the hindered rotation may be expressed by the modified Fourier series,

$$V(\alpha) = \frac{V_3}{2}(1 - \cos 3\alpha) + \frac{V_6}{2}(1 - \cos 6\alpha) + \ldots \tag{3}$$

where V_3 is the height of the three-fold potential barrier, V_6 of the six-fold one, and so on. In most cases, the V_6 and higher terms may be dropped (Fateley and Miller 1961) since V_6 has been shown to be much less than V_3.

Clearly, the problem of determining the form of the potential barrier (for example the potential equation) is of prime importance and will vary as a function of the symmetry of the molecule. In the general case, it is hoped that sufficient experimental data will be available to evaluate the appropriate coefficients (for example the barrier heights) V_1, V_2 and V_3 in eqn (1), or simply V_3 in eqn (3). For the case of internal rotation about the C–C single bond (eqn 3), three pieces of input data are necessary for evaluation of V_1, V_2 and V_3. Unfortunately, for most molecules only one experimental datum is available—the frequency of the $0 \rightarrow 1$ transition in the central potential well. It is, therefore, often necessary to make simplifying assumptions. This problem will be discussed in more detail in Section IIIC.

A. The Three-fold Barrier

In the previous section, eqn (3) represented the expression for the potential energy hindering rotation about a three-fold symmetry axis. As the three-fold barrier represents one of the simpler cases of barrier calculations, and has been the most well-developed experimentally, it will be presented here in some detail.

As pointed out earlier, all terms after the first may be dropped (see later discussion concerning the significance of the V_6 term) leaving the three-fold barrier potential in the form:

$$V(\alpha) = \frac{V_3}{2}(1 - \cos 3\alpha). \tag{4}$$

When this potential function is substituted in the wave equation for torsional motion, a function in the form of Mathieu's equation is obtained.† The energy solution for each torsional quantum number v gives two sublevels: one is non-degenerate (A) and the other is doubly

† Mathieu's equation may be written in the form: $y'' + (b - S\cos^2 x)y = 0$. For a fixed value of S, there exists an infinite sequence of characteristic b values, corresponding to which the solutions y are periodic and of period π or 2π. Blanch and Rhodes (1965) have presented these in tables as a function of t, where $t = 1/\sqrt{S}$. For calculations presented in this book, these values have been recalculated to express b in terms of S (for a complete tabulation of these values, see Section II of AFML-TR-66-408, January, 1967 Air Force Report). In order to be consistent with Blanch and Rhodes' notation, R is used for v, the principal torsional quantum number.

degenerate (E). The torsional energy is given by (see Herschbach 1959 for notation):

$$E_{v\sigma} = \left(\frac{n}{2}\right)^2 Fb_{v\sigma} = \dagger 2.25 Fb_{v\sigma} \tag{5}$$

where n = barrier symmetry (i.e. three-fold, etc.);
v = principle torsional quantum number;
σ = an index designating either the A- or E-level;
$b_{v\sigma}$ = an eigenvalue of the Mathieu equation;
$F = h^2/8\pi^2 rI_\alpha$;
rI_a = reduced moment of inertia for internal rotation
$= I_a\left[1 - I_a \sum_g (\lambda_g^2/I_g)\right]$;
I_a = moment of inertia of the internal top (i.e. CH_3) about its own symmetry axis;
λ_g = cosine of the angle between the axis of the internal top and the gth principal axis of inertia of the entire molecule;
I_g = gth principal moment of inertia of the entire molecule.

The observed infrared frequency gives the difference between two of the energy levels which according to eqn (5) is

$$\Delta E = 2.25 F(\Delta b_{v\sigma}). \tag{6}$$

Thus if the absorption frequency is known (torsion), and sufficient structural information‡ is available to determine F, $\Delta b_{v\sigma}$ can be calculated. From $\Delta b_{v\sigma}$ a dimensionless parameter S can be obtained from solutions for the Mathieu equation. The barrier height can then be calculated from S by using:

$$V_3 = 2.25 FS. \tag{7}$$

Herschbach (1957) has prepared tables of $b_{v\sigma}$ *versus* S for a series of v values. Graphs of $\Delta b_{v\sigma}$ *versus* S (Fig. 2) have been produced using this data, and will be found useful in calculations.

Acetaldehyde represents a particularly favourable case of a single methyl group bonded to a "rigid" framework in which there are three identical potential minima. The theory for this molecule and other similar molecules has been well developed by Wilson (see Kilb *et al.* 1957) through microwave studies. In addition, Lin and Swalen (1959) have

† For three-fold barrier.

‡ Obtaining such structural information may be a formidable task. In many cases this information may be found in the literature, for example Herschbach (1959), Kilb *et al.* (1957). Mathematical treatments for the calculation of reduced moments of inertia may be found in Pitzer (1942, 1946), Lin and Swalen (1959), and Quade and Lin (1963).

published an excellent review in which the theory and many results have been presented. It is valuable in the case of acetaldehyde to have microwave data available in that infrared data represents a completely

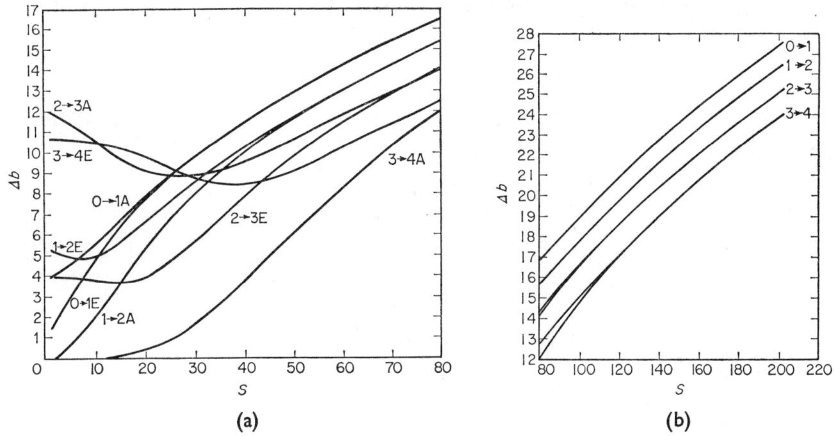

FIG. 2. Graphs of $\Delta b_{v\sigma}$ *versus* S for various transitions, assuming a three-fold potential barrier.

independent check. Fateley and Miller (1961) have assigned the torsion in acetaldehyde to a band at 150 cm^{-1} (Fig. 3). It may be noted that in this region three absorptions are observed which are assigned to rota-

TABLE II. Infrared Bands and Potential Barriers of CH_3CHO.

Compound (gas phase)	Observed (cm^{-1})	Intensity	Pressure in 1-m cell (cm Hg)	Assignment of CH_3 torsion	V_3 (cal mole^{-1}) Infrared	Microwave
CH_3CHO	150 ± 1	*vw*	(solid)	$0 \to 1$ (A)	—	—
	143	*s*				
	150	*m*	8	$0 \to 1$ (A)	1180	1168 ± 30[a]
	158	*m*				

[a] Herschbach (1959).

tional structure, irregularly spaced due to coupling between internal and overall rotation.

If the 150 cm^{-1} band is assumed to be the $0 \to 1$ transition of the methyl torsion and, in addition, it is assumed that the A and E levels cannot be resolved, then using the A levels

$$\Delta E_{v\sigma} = E_{1A} - E_{0A} = 150 \text{ cm}^{-1} = 2.25F(\Delta b_{v\sigma}). \tag{8}$$

The value of F for acetaldehyde (Herschbach 1959) is 230.7 kMc, or 7.696 cm^{-1}.

$$\Delta b_{v\sigma} = b_{1A} - b_{0A} = 8.663. \tag{9}$$

The value of S is then calculated from interpolation of the tables of Δb *versus* S with the result that

$$S = 23.84.$$

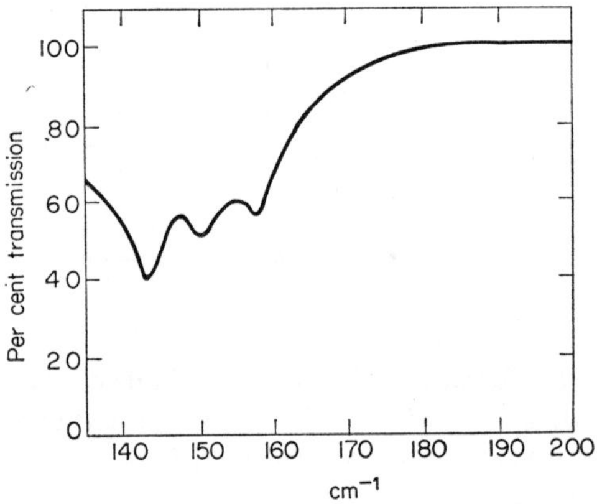

FIG. 3. Infrared absorption of acetaldehyde vapour at low frequencies. Path length = 1 m. (a) 8 cm pressure; slit width = 8.5 cm^{-1}. (b) 40 cm pressure; slit width = 3.5 cm^{-1}. There was no other absorption between 80 and 350 cm^{-1}.

Substitution of this value in eqn 7 yields

$$V_3 = 413 \text{ cm}^{-1} = 1180 \text{ cal mole}^{-1}.$$

Microwave data, based on the same theory of energy levels but on a different set of transitions, gives a value of 1168 ± 30 cal mole^{-1} which is in excellent agreement with Herschbach (1959).

By reversing the calculations, the value for the barrier can be used to determine the energies of the A and E sublevels for various torsional quantum numbers. These values are given in Table III and are drawn to scale in Fig. 4. Obviously, the splitting of the A and E levels for $v = 0$ (0.09 cm^{-1}) and $v = 1$ (2.2 cm^{-1}) could not be observed, justifying the earlier assumption.

Absorption bands in the spectrum of acetaldehyde have also been

TABLE III. Energy Levels for Acetaldehyde. ($V_3 = 412.7\ \text{cm}^{-1} = 1180\ \text{cal mole}^{-1}$).

v	Sublevel	Energy (cm^{-1})	Δ (cm^{-1})
0	A	79.58	0.09
	E	79.67	
1	E	227.4	2.2
	A	229.6	
2	A	340.6	15.3
	E	355.9	
(Barrier)		(412.7)	
3	E	435.8	62.0
	A	497.8	
4	A	513.3	86.1
	E	599.4	

observed between 257 cm^{-1} and 280 cm^{-1} and have been attributed by Fateley and Miller (1961) to the $0 \rightarrow 2$ transition, for which the absorption band is expected to be complex due to a 15 cm^{-1} splitting of the A and E levels as well as coupling of internal and overall rotation. Further investigation of this region showed the presence of the trimer of acetaldehyde complicating the spectrum, therefore an unequivocal assignment of these overtone bands was not possible.

It should be noted that the height of the barrier is quite sensitive to

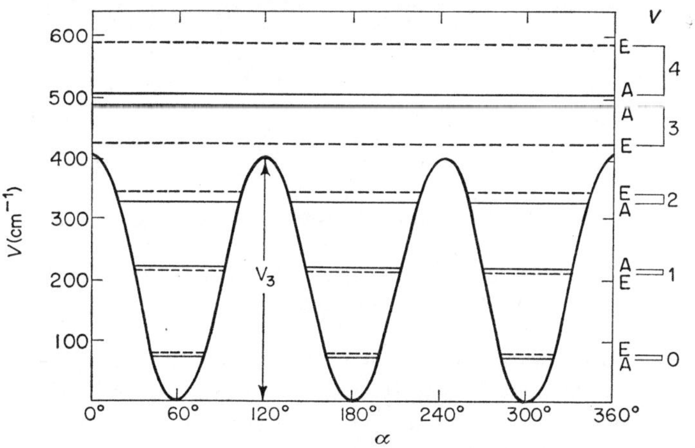

FIG. 4. Potential function and energy levels for hindered rotation. $V(\alpha) = V_3(1 + \cos 3\alpha)/2$. Drawn to scale for acetaldehyde, with $V_3 = 412.7\ \text{cm}^{-1} = 1180$ cal/mole. $V_3 = 408\ \text{cm}^{-1}$, $F = 7.090\ \text{cm}^{-1}$.

the exact value of the infrared frequency. For example, in the case of acetaldehyde, lowering the observed frequency by 1 cm^{-1} lowers the barrier by 15 cal $mole^{-1}$. In addition, the barrier heights depend on the

TABLE IV. Torsional Frequencies and Barriers to Internal Rotation for Some Substituted Toluenes.

	ν (cm^{-1}) Torsion	F (cm^{-1})	S	V_3 (kcal $mole^{-1}$)
o-Fluorotoluene	173	5.437	57.93	2.03
o-Chlorotoluene	150	5.437	44.65	1.56
o-Methyltoluene (Xylene)	180	5.437	62.38	2.18

structure parameter F. Changing F for acetaldehyde from 7.696 to 7.896 (2.6%) changes V_3 from 1180 to 1154 cal $mole^{-1}$ (2.5%).

Another excellent example of a three-fold barrier is provided by a series of *ortho*-substituted toluenes studied by Fateley and co-workers (private communication). Table IV gives the assigned torsional frequen-

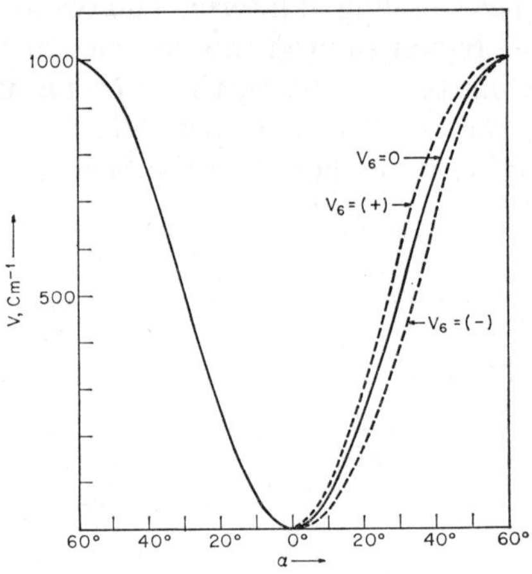

FIG. 5. Affect of V_6 on the shape of the potential curve. $V_3 = 1{,}000$, $V_6/V_3 = -0.1$, 0, and +0.1. $V(\alpha) = V_3(1 - \cos 3\alpha)/2 + V_6(1 - \cos 6\alpha)/2$.

cies, the parameters F and S for eqn (7), and the calculated barriers (V_3). The calculations are exactly analogous to those previously described for acetaldehyde. *Meta*- and *para*-substituted toluenes exhibit no absorptions down to 50 cm^{-1} which can be assigned to the torsion. This is consistent with the observation that six-fold barriers are expected to be quite low. Indeed, several molecules containing methyl groups with effectively six-fold symmetry (for example CH_3NO_2, CH_3BF_2, etc) exhibit nearly free rotation.

The addition of the V_6 term to the potential function does not change the height of the three-fold barrier because the V_6 term makes no contribution at either the potential minimum or maximum. The barrier height, therefore, is still given by V_3. The V_6 term does, however, change the shape of the potential well; this is shown in an exaggerated way in Fig. 5. A positive V_6 makes the well narrower and the maximum broader. It therefore causes the torsional levels for $v = 0, 1, 2...$ to be separated to a greater degree than they would if V_6 were zero. A negative V_6 has the opposite effect, that is the levels are slightly compressed relative to their spacings for $V_6 = 0$.

B. The Two-fold Barrier

Phenol represents an excellent example of a molecule with a simple two-fold barrier. Fateley and co-workers (unpublished results) have observed two frequencies in the far infrared (Fig. 6) at 310 cm^{-1} and

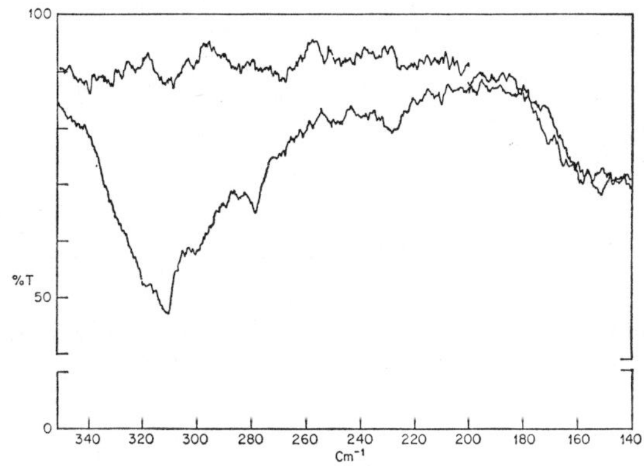

FIG. 6. Phenol, vapour, vapour pressure at 23°C, path length 8.2 m, polyethylene windows.

279 cm^{-1}, the former being assigned to the $0 \rightarrow 1$ torsional transition and the latter as a "hot band" representing the average of the $1 \rightarrow 2$ transitions. The $2 \rightarrow 3$ transition is predicted at 258 cm^{-1}, but was not observed.

The theory for the calculation of the barrier height as applied to the two-fold barrier would yield the equation for the potential as follows

$$V(\alpha) = \frac{V_2}{2}(1 - \cos 2\alpha) + \frac{V_4}{2}(1 - \cos 4\alpha) \tag{10}$$

where V_2 represents the height of the two-fold barrier and V_4 that of the four-fold one. Following the derivation as outlined in the preceding section, the barrier height is given by

$$V_2 = FS. \tag{11}$$

Again, solving the $\Delta b_{v\sigma}$ from the assigned frequency gives a value of S from Herschbach's tables. The value of F in cm^{-1} calculated by standard procedures is 25.6809 for phenol. Substitution in eqn 11 yields a barrier height, V_2, of 1143 cm^{-1}. V_2 was also calculated using the 279 cm^{-1} "hot band". Since this represents an average of the $1 \rightarrow 2$ transitions,

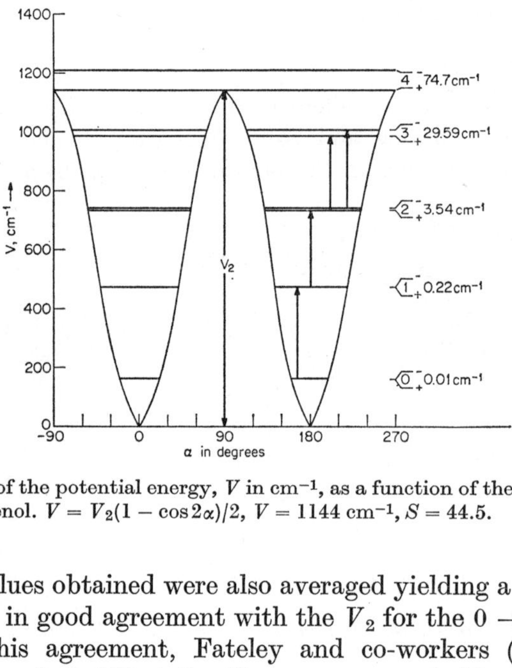

FIG. 7. A plot of the potential energy, V in cm^{-1}, as a function of the internal angle of rotation α for phenol. $V = V_2(1 - \cos 2\alpha)/2$, $V = 1144$ cm^{-1}, $S = 44.5$.

the two V_2 values obtained were also averaged yielding a value of 1140 cm^{-1} which is in good agreement with the V_2 for the $0 \rightarrow 1$ transition. Because of this agreement, Fateley and co-workers (private communication) concluded that the V_4 term was probably equal to zero. The potential energy plotted as a function of internal angle, α, for phenol is given in Fig. 7.

The V_4 term can often be calculated in cases where asymmetries are introduced in the molecule. In the case of *ortho*- or *meta*-substituted phenols, the barrier becomes slightly unsymmetrical and the V_4 term non-zero.

V_4 has been calculated by Fateley and co-workers (private communication) for *trans meta*-chlorophenol and found to be -6.98 cm^{-1} or approximately 0.5% of V_2 (1163 cm^{-1}).

C. Unsymmetrical Barriers

In previous sections, discussion has been confined to barrier calculations based on potential functions which were radically simplified (for example three-fold and two-fold barriers) due to inherent molecular symmetry. Many molecules, however, (the majority, unfortunately) do not possess such symmetry and therefore present a more formidable problem with respect to barrier calculations. Equations (1) and (2) and Fig. 1 represent the form of the potential in the most general sense and as mentioned earlier the problem rests with the evaluation of the individual V terms. In most cases, a deficiency of experimental data exists (for example usually only the $0 \rightarrow 1$ transition is observable). Clearly, several assumptions and approximations must be made in order to pursue the unsymmetrical barrier calculations.

Although each molecular system must be considered independently, Fateley and co-workers (1965) provide an excellent example in their studies of torsions around –C–C– single bonds in molecules containing the group =C–C= (for example butadiene, glyoxal, etc). In these molecules, certain predictions can be made regarding the relative significance of the V_1, V_2 and V_3 terms in eqn 1. First, the concept of resonance leads to the observation that the planar forms

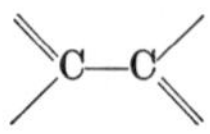

trans

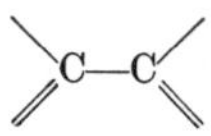

cis

will be more stable than any of the nonplanar conformations. This suggests that V_2 will be of greater significance since it is the only term with minima at both $\alpha = 0$ and $180°$. There is also evidence that the *trans* form is more stable than the *cis* (see Harris and Witkowski 1964). Since the V_2 does not account for this asymmetry, the V_1 and V_3 terms must be considered and since it is impossible to determine *a priori* which is more significant all three terms must be used.

If the bonding in these systems is considered using Pauling's view (Pauling 1960) of a double bond as two bent single bonds, then the potential barrier is still essentially trigonal which justifies the inclusion of the $\cos 3\alpha$ term in eqn 1. In addition, it provides a basis for rationalizing the *cis-trans* energy difference. The *trans* form is more stable than the *cis* because it is a staggered conformation, whereas the *cis* is an eclipsed one. Furthermore, it predicts that, for this term, the metastable conformation is not *cis*, but a pair of gauche forms with $\alpha = \pm 120°$.

The actual metastable conformation will be determined by the sum of V_1, V_2 and V_3. If one differentiates eqn (1), it can be shown that the *cis* form ($\alpha = \pm\pi$) will always be at a potential maximum or minimum. It will be a minimum if $(-V_1 + 4V_2 - 9V_3)$ is positive, and a maximum if it is negative. Therefore, the metastable form will be *cis* if $4V_2 > V_1 + 9V_3$ or nonplanar if $4V_2 < V_1 + 9V_3$.

In the case where there is only one experimental datum, the calculation proceeds as follows.

A quantity V^* is defined as

$$V^* = V_1 + 4V_2 + 9V_3 \tag{12}$$

and

$$F\ (\text{in cm}^{-1}) = h/8\pi^2 cI_r = \frac{16.825}{I_r\ (\text{amu-A}^2)} \tag{13}$$

where F has the form of a rotational constant; I_r = reduced moment of inertia for the torsion about the =C–C= bond (see Pitzer 1946); $\tilde{\nu}$ = observed infrared transition in cm^{-1}.

The simplifying assumption that the torsional oscillation is harmonic may be made and $\cos\alpha$ in eqn (1) expanded as a power series in α, terminating after α^4,

$$V = \tfrac{1}{4}(V_1 + 4V_2 + 9V_3)\,\alpha^2 - \tfrac{1}{48}(V_1 + 16V_2 + 81V_3)\,\alpha^4. \tag{14}$$

If α is small, the last term may be dropped, yielding

$$V = \tfrac{1}{4}(V_1 + 4V_2 + 9V_3)\,\alpha^2 = \tfrac{1}{4}V^*\alpha^2. \tag{15}$$

The equation for V is now in the form of a harmonic oscillator (for example $V(x) = \frac{1}{2}Kx^2$). In effect, the curvature at the minimum of the parabolic potential curve has been matched to that of a true harmonic oscillator with a cosine-based function. In doing this, the barrier height, V^*, is introduced into a harmonic oscillator.

It is next necessary to relate V^* to the observed experimental datum, the $0 \rightarrow 1$ torsional transition. Using the harmonic oscillator results, with the effective mass being I_r and the force constant $\frac{1}{2}V^*$, the expression

$$V^*\ (\text{ergs}) = 8\pi^2 c^2 \tilde{\nu}^2 I_r \tag{16}$$

is obtained, which when divided through by hc to convert V^* to cm^{-1}, and introducing F, yields

$$V^*\ (\text{cm}^{-1}) = \frac{\tilde{\nu}^2}{F} = V_1 + 4V_2 + 9V_3. \tag{17}$$

Thus from a knowledge of the torsional fundamental $\tilde{\nu}$, and sufficient molecular geometry to obtain F, V^* can be evaluated.

It is of interest to indicate qualitatively why the coefficients 4 and 9 appear in the definition of V^*. The force constant for this harmonic oscillator is $\frac{1}{2}V^*$ and it is also the curvature at $\alpha = 0$. Figure 1 shows that for equal values of V_n, the curvature is greater for the V_3 than the V_2 term, and similarly that is greater than the V_1 term. Hence, the coefficients for V_3, V_2 and V_1 decrease in that order. The corollary to this is that a relatively small V_3 may be important in determining the observed frequency because it has a relatively large coefficient.

With only one experimental datum, V^* is probably the closest estimate of the barrier value which may be obtained. However, this value does hold a significant amount of information. Table V lists the compounds

TABLE V. Observed Torsional Frequencies ($\tilde{\nu}$), Reduced Moments of Inertia (I_r), F Values, and Barrier Heights (V^*).

Compound	$\tilde{\nu}$ (cm^{-1})	I_r (amu-A^2)	F (cm^{-1})	$V^* = V_1 + 4V_2 + 9V_3$ (cm^{-1})	(kcal $mole^{-1}$)
Butadiene	163	6.228	2.706	9820	28.1
Isoprene	153	8.989	1.875	12480	35.7
2,3-Dimethylbutadiene	not found	27.97	0.602	—	—
Acrolein gas	157	5.080	3.317	7430	21.2
liquid	182	5.080	3.317	9990	28.6
Methyl vinyl ketone	101 ± 2	9.074	1.857	5490	15.7
Methacrolein	165	6.905	2.441	11150	31.9
Methyl isopropenyl ketone	not found	—	—	—	—
Glyoxal	128	4.949	3.405	4810	13.7
Pyruvaldehyde	105 ± 2	6.634	2.54	4340	12.4
Biacetyl	48	25.71	0.655	3520	10.1

studied by Fateley *et al.* (1965) including the torsional frequencies, I_r values, F and V^*. Referring to Fig. 8, several of these molecules have been arranged to illustrate certain structural differences affecting the value of V^*. For example, the structural change upon going down the columns is to replace a $=CH_2$ group by an $=O$ atom. In all cases, each such substitution lowers V^*. This is reported to be the result of the partial double bond nature of the central C–C bond and since oxygen is much more electronegative than $=CH_2$, there will be less π orbital overlap when the $=CH_2$ is replaced by oxygen, thus reducing the barrier. In going across the rows (left to right) the structural change is to replace the central hydrogens by methyl groups. V^* appears to increase when the methyl group is substituted on the vinyl group, while it decreases when substituted on the aldehyde group. No interpretation has been offered for this latter observation.

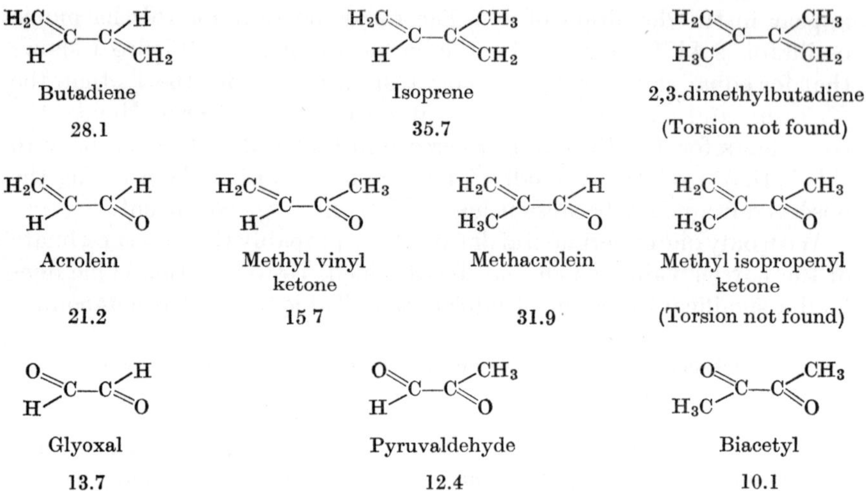

FIG. 8. Measured values of $V^* = V_1 + 4V_2 + 9V_3$ in kcal mole^{-1} for a number of molecules with unsymmetrical barriers.

Evaluation of V_1, V_2 and V_3 separately is only possible if more experimental data is available such as hot bands or overtones and in some cases from other data such as electronic spectra or microwave data. Details of these calculations may be found in the work of Fateley *et al.* (1965) applied to acrolein for which torsional overtones obtained from electronic spectra were utilized.

An additional interesting study by Miller *et al.* (1967) concerns torsions about –C–C– single bonds in benzaldehyde and substituted benzaldehydes. Again, lack of sufficient experimental data prohibits the evaluation of all the V terms independently; however, utilizing the methods outlined above and the assumption that the potential function in the region of the torsion is harmonic, eqn (17) is taken to be valid, for example

$$V^* = \frac{\tilde{\nu}^2}{F}$$

where $V^* = V_1 + 4V_2 + 9V_3$.

In the case of asymmetrically-substituted benzaldehydes such as *ortho*-chlorobenzaldehyde

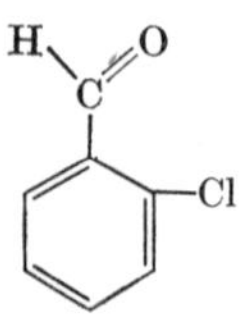

one can make certain qualitative predictions concerning the relative importance of the V_1, V_2 and V_3 terms. The factors giving rise to the V_1 term must have a periodicity of 2π in α, steric interaction being an obvious possibility. In addition, there may be a small contribution from hydrogen bonding between the halogen atom and the aldehydic hydrogen atom, the effect of which will be most important in the *ortho* derivatives, smaller in the *meta* ones, and zero for the *para* ones. The two-fold barrier, V_2, arises mainly from overlap between the π orbitals of the carbonyl bond with those of the aromatic ring. Consequently, this coefficient would be expected to be relatively important, regardless of the substitution pattern. The three-fold barrier, V_3, is judged to be insignificant since it is difficult to conceive any reason for one to exist; for example, the V_3 term for acrolein (Fateley *et al.* 1965) was found to be very small.

If the assumption is made that $V_3 = 0$ then

$$V^* = V_1 + 4V_2. \tag{18}$$

In the case of symmetrically-substituted benzaldehydes (for example *para* derivatives and 3,5- or 2,6-disubstituted derivatives) $V_1 = V_3 = 0$ and

$$V^* = 4V_2 = \frac{\tilde{\nu}^2}{F} \tag{19}$$

$$V_2 = \frac{\tilde{\nu}^2}{4F}. \tag{20}$$

F values for the benzaldehydes studied were obtained using molecular dimensions developed by Quade (1962) from microwave studies. At best, however, the calculated F values are "educated estimates" since certain assumptions have been made about the geometry of the molecules. Calculations based on F values for the *ortho* and *meta* isomers have an added complication in that they may exist in two planar conformations which are more stable than the nonplanar ones. These conformations, *cis* and *trans*, have different F values.

Calculations of the barrier for unsubstituted and symmetrically-substituted benzaldehydes are relatively straightforward. Barrier values obtained in this manner have been previously given in Table V.

For the *meta*-substituted series, Miller *et al.* (1967) observed two low-frequency absorptions separated by 6–16 cm^{-1}. After careful consideration of all possible assignments they concluded that these bands most likely represent two rotational isomers co-existing in appreciable concentrations and having slightly different torsional frequencies. The best support for this conclusion comes from the fact that the *meta* derivatives were the only ones to have such bands and they are the most likely to exist as *cis-trans* isomers. For the *ortho* isomers, the steric effect is likely to be quite large so that only one of the planar forms would have an appreciable population. In the *meta* isomer, the steric effect is much smaller and the two rotamers are more likely to have similar energies so that each has sufficient population to be observed. Even if their energies were identical, their torsional frequencies would be different because they have different F values. For the mathematical treatment of the *meta* case, the reader is referred to the original literature. The results, however, of the above described assumption are presented in Fig. 9. It can be seen that two potential minima are observed and that the most stable form is predicted to be the *o-cis*.

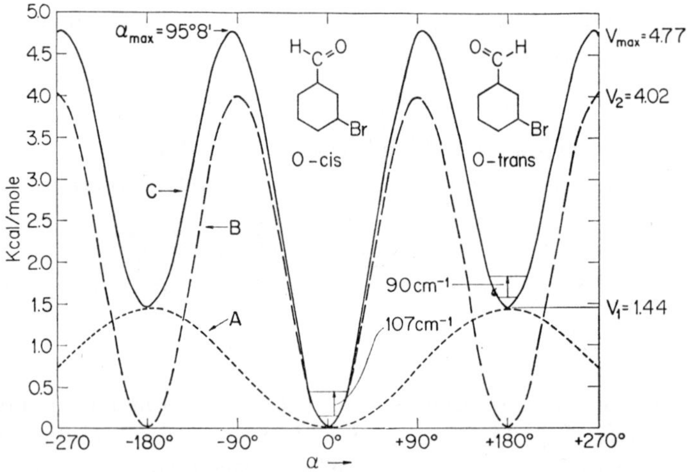

FIG. 9. Potential curve for the torsion in m-Br-benzaldehyde (to scale). $V(\alpha)$ (kca mole^{-1}) $= 1.44(1 - \cos\alpha)/2 + 4.02(1 - \cos 2\alpha)/2$. (a) First term. (b) Second term. (c) Sum. V_3 is assumed to be zero.

For the *ortho*-substituted series, the approach is similar to the above. Two absorptions were observed for both the *o*-fluoro and *o*-chlorobenzaldehyde. Miller and co-workers have suggested that these bands may be due to rotamers resulting from steric effects and that their computations indicate that the *o-trans* configuration is more stable. Steric considerations favour the form with the oxygen turned away from the halogen and, in addition, a weak intramolecular hydrogen bond may also stabilize this form. For *o*-bromo- and *o*-methyl-benzaldehyde only one torsional frequency was observed which perhaps is a result of the size of the substituent. The *o-cis* form would be relatively unstable. Assuming the *o-trans* to be more stable, a $V^* = V_1 + V_4$ is the best that could be calculated.

References

Blanch, G. and Rhodes, I. (1965). *J. Wash. Acad. Sci.* **188**.
Fateley, W. G., Matsubara, I., and Witkowski, R. E. (1964). *Spectrochim. Acta* **20**, 1461.
Fateley, W. G. and Miller, F. A. (1961). *Spectrochim. Acta* **17**, 857.
Fateley, W. G. and Miller, F. A. (1963). *Spectrochim. Acta* **19**, 611.
Fateley, W. G., Harris, R. D., Miller, F. A., and Witkowski, R. E. (1965). *Spectrochim. Acta* **21**, 231.
Harris, R. K. and Witkowski, R. E. (1964). *Spectrochim. Acta* **20**, 1651.
Herschbach, D. R. (1957). *J. Chem. Phys.* **27**, 975.
Herschbach, D. R. (1959). *J. Chem. Phys.* **31**, 91.
Kilb, R. W., Lin, C. C., and Wilson, Jr., E. B. (1957). *J. Chem. Phys.* **26**, 1695.
Lin, C. C. and Swalen, J. D. (1959). *Revs. mod. Phys.* **31**, 841.
Miller, F. A., Fateley, W. G., and Witkowski, R. E. (1967). *Spectrochim. Acta* **23A**, 891.
Pauling, L. (1960). "The Nature of the Chemical Bond." 3rd Edition, pp. 130–142, 292. Cornell University Press, Ithica, New York.
Pitzer, K. S. (1942). *J. Chem. Phys.* **10**, 605.
Pitzer, K. S. (1946). *J. Chem. Phys.* **14**, 239.
Quade, C. R. (1962). *Diss Abstr.* **23**, 1750.
Quade, C. R. and Lin, C. C. (1963). *J. Chem. Phys.* **38**, 540.
Radcliffe, K. and Wood, J. L. (1966a). *Trans. Faraday Soc.* **62**, 1678.
Radcliffe, K. and Wood, J. L. (1966b). *Trans. Faraday Soc.* **62**, 2038.

APPENDIX

Measured Values for Barriers to Internal Rotation

A. Rotational Barriers for H-Substituted Ethanes

Compound	Formula	Barrier (kcal-mole^{-1})	References
Ethane	CH_3CH_3	2.875 ± 0.125	38
Ethane	CH_3CH_3	2.88	34
Ethane	CH_3CH_3	2.3	21
Ethane	CH_3CH_3	2.85	1
Ethane	CH_3CH_3	2.875	24
Ethane	CH_3CH_3	1.77 and 2.63	9
Ethane	CH_3CH_3	3.3	56
Ethane	CH_3CH_3	2.4	70
Ethane	CH_3CH_3	0.7–2.1	18
1-Monofluoroethane	CH_3CH_2F	$3.30 \pm .03$	8
1-Monofluoroethane	CH_3CH_2F	$3.31 \pm .21$	67
1-Monofluoroethane	CH_3CH_2F	$3.306 \pm .100$	67
1-Monofluoroethane	CH_3CH_2F	3.30	34
1-Monofluoroethane	CH_3CH_2F	3.30	20
1,1-Difluoroethane	CHF_2CH_3	3.20	34
1,1-Difluoroethane	CHF_2CH_3	3.18	20
1,1,1-Trifluoroethane	CF_3CH_3	3.48	34

A. Rotational Barriers for H-Substituted Ethanes—*continued*

Compound	Formula	Barrier (kcal-mole^{-1})	References
1,1,1,2-Tetrafluoroethane	CF_3CH_2F	4.2	12
1,1,1,2-Tetrafluoroethane	CF_3CH_2F	4.58	69
1,1,1,2,2-Pentafluoroethane	CF_3CF_2H	3.35	6
1,1,1,2,2,2-Hexafluoroethane	CF_3CF_3	10.8	44
1-Monochloroethane	CH_2ClCH_3	$3.560 \pm .012$	39
1-Monochloroethane	CH_2ClCH_3	3.56	34
1-Monochloroethane	CH_2ClCH_3	3.69	60
1,1,1,2,2,2-Hexachloroethane	CCl_3CCl_3	10.8	44
1-Monobromoethane	CH_2BrCH_3	$3.567 \pm .030$	39
1-Monobromoethane	CH_2BrCH_3	3.56	34
1-Monobromoethane	CH_2BrCH_3	3.57	40
1-Chloro-1,1-difluoroethane	CF_2ClCH_3	3.94	6
1-Chloro-1,1,2,2,2-pentafluoroethane	CF_2ClCF_3	5.67	59
1-Chloro-1,1,2,2,2-pentafluoroethane	CF_2ClCF_3	5.10	6
1,2-Dichloro-1,1,2,2-tetrafluoroethane	CF_2ClCF_2Cl	5.27	6
1,2 Difluoro-1,1,2,2-tetrachloroethane	CCl_2FCCl_2F	9.65	59
1-Chloro-2,2,2-trifluoroethane	CH_2ClCF_3	5.79	69
1-Bromo-2,2,2-trifluoroethane	CH_2BrCF_3	6.21	69
1-Bromo-2,2,2-trifluoroethane	CH_2BrCF_3	5.52	6
1-Bromo-1,1,2,2,2-pentafluoroethane	CF_2BrCF_3	6.40	59
1-Iodoethane	CH_2ICH_3	3.22	25
1-Iodo-2,2,2-trifluoroethane	CH_2ICF_3	6.13	69
1-Iodo-1,1,2,2,2-pentafluoroethane	CF_2ICF_3	7.09	59
Propyl nitrile	CH_3CH_2CN	3.05	34

B. Rotational Barriers in Carbonyl Compounds

Compound	Formula	Barrier (kcal-mole^{-1})	References
Acetaldehyde	CH_3CHO	$1.103 \pm .060$	67
Acetaldehyde	CH_3CHO	$1.162 \pm .030$	67
Acetyl fluoride	CH_3CFO	1.041	52
Acetyl fluoride-C^{13}	$C^{13}H_3CFO$	1.041	52
Acetyl fluoride-O^{18}	CH_3CFO^{18}	1.055	52
Acetyl fluoride-d_3	CD_3CFO	1.031	52
Acetyl chloride	CH_3COCl	$1.296 \pm .030$	19
Acetyl bromide	CH_3COBr	$1.305 \pm .030$	32
Acetyl iodide	CH_3COI	$1.301 \pm .030$	6
Acetyl cyanide	CH_3COCN	$1.210 \pm .030$	31
Trifluoroacetaldehyde	CF_3CHO	0.95	43

B. Rotational Barriers in Carbonyl Compounds—*continued*

Compound	Formula	Barrier (kcal-mole^{-1})	References
Tetrafluoroacetaldehyde	CF_3COF	1.39	43
Dimethyl ketone	$(CH_3)_2CO$	0.78	64
Dimethyl ketone-d$_3$	$(CD_3)_2CO$	0.738	64
Dimethyl ketone	$(CH_3)_2CO$	0.778	45
Dimethyl ketone-C^{13}	$(C^{13}H_3)_2CO$	0.777	45
Dimethyl ketone-d$_3$	$(CD_3)_2CO$	0.726	45
Methyl formate	$HCOOCH_3$	$1.190 \pm .040$	11
Formamide	$HCONH_2$	$1.058 \pm .143$	10
1-Butyn-3-one	$HC{\equiv}CCOCH_3$	1.07	62
1-Buten-3-one	$H_2C{=}CHCOCH_3$	1.25	17

C. Rotational Barriers for C- and H-Substituted Propanes, Propenes, Propynes and Related Systems

Compound	Formula	Barrier (kcal-mole^{-1})	References
Propane	$CH_3CH_2CH_3$	3.3	20
Propane	$CH_3CH_2CH_3$	>2.7	41
trans n-Propyl fluoride	$CH_3CH_2CH_2F$	2.69	22
gauche n-Propyl fluoride	$CH_3CH_2CH_2F$	2.87	22
trans n-Propyl chloride	$CH_3CH_2CH_2Cl$	2.78	58
gauche n-Propyl chloride	$CH_3CH_2CH_2Cl$	2.96	58
trans n-Propyl bromide	$CH_3CH_2CH_2Br$	2.36	58
gauche n-Propyl bromide	$CH_3CH_2CH_2Br$	2.65	58
Propene	$CH_3CH{:}CH_2$	$1.978 \pm .017$	42
trans-1-Fluoro-1-propene	$CH_3CH{:}CHF$	$2.2 \pm .10$	61
cis-1-Fluoro-1-propene	$CH_3CH{:}CHF$	1.06	5
trans-1-Chloro-1-propene	$CH_3CH{:}CHCl$	$2.17 \pm .10$	2
cis-1-Chloro-1-propene	$CH_3CH{:}CHCl$	0.62	4
2-Fluoro-1-propene	$CH_3CF{:}CH_2$	$2.44 \pm .016$	53
2-Chloro-1-propene	$CH_3CCl{:}CH_2$	$2.671 \pm .02$	66
3-Fluoro-1-propene	$CH_2FCH{:}CH_2$	~ 2.0	23
2-Methyl-1-propene	$(CH_3)_2C{:}CH_2$	2.21	35
trans-2-Butene	$CH_3CH{:}CHCH_3$	1.95	13
cis-2-Butene	$CH_3CH{:}CHCH_3$	0.75	4
trans-2-Butenenitrile	$CH_3CH{:}CHCN$	>2.10	51
cis-2-Butenenitrile	$CH_3CH{:}CHCN$	1.40	3
cis-1-Chloro-2-fluoro-1-propene	$CH_3CF{:}CHCl$	$2.565 \pm .03$	63
trans-1-Propene-3-imine	$CH_2{:}CHCH{:}NH$	$2.608 \pm .026$	37
2-Butyne	$CH_3C{\vdots}CCH_3$	.10	27
2-Butyne	$CH_3C{\vdots}CCH_3$	.03	7
2-Butyne	$CH_7C{\vdots}CCH_3$	$<.10$	30
1-Chloro-2-butyne	$CH_3C{\vdots}CCH_2Cl$	0.1	36
1,4-Dichloro-2-butyne	$ClCH_2C{\vdots}CCH_2Cl$	0.29	1

D. Rotational Barriers for Silicon and Germanium Compounds

Compound	Formula	Barrier (kcal-mole^{-1})	References
Methyl silane	CH_3SiH_3	1.670	24
Methyl silane	CH_3SiH_3	1.70	26
Dimethyl silane	$(CH_3)_2SiH_2$	1.665 ± .010	49
Dimethyl silane	$(CH_3)_2SiH_2$	1.65	50
Trimethyl silane	$(CH_3)_3SiH$	1.4–2.0	54
Methyl monofluorosilane	CH_3SiH_2F	1.56	48
Methyl monofluorosilane	CH_3SiH_2F	1.559 ± .030	47
Methyl difluorosilane	CH_3SiHF_2	1.28	65
Chloromethylsilane	CH_2ClSiH_3	2.55	60
Trichloromethyltrichlorosilane	CCl_3SiCl_3	4.3	44
Hexachlorodisilane	$SiCl_3SiCl_3$	1.0	44
Methyl germane	CH_3GeH_3	1.239 ± .025	33
Methyl germane	CH_3GeH_3	1.239	24

E. Rotational Barriers for Miscellaneous Aliphatic Compounds

Compound	Formula	Barrier (kcal-mole^{-1})	References
Methanol	CH_3OH	1.07	24
Methanol-d_1	CH_2DOH	1.17	57
Methanethiol	CH_3SH	1.27	57
Methanethiol	CH_3SH	1.26	28
Methylamine	CH_3NH_2	1.976	57
Methylamine	CH_3NH_2	1.94	46
Methyl phosphine	CH_3PH_2	1.959	57
Methyl phosphine	CH_3PH_2	1.96	29
Nitrosomethane	CH_3NO	2.69 ± 0.57	14
Methyl nitrite	CH_3NO_2	0.006	55
Methyl nitrate	CH_3NO_3	2.32	15
Methyl formate	CH_3O_2CH	1.19 ± .04	11
Trimethyl amine	$(CH_3)_3N$	4.41	16
Trimethyl arsine	$(CH_3)_3As$	1.5–2.5	40
Trifluoromethyl phosphine	CF_3PH_2	2.36 ± .08	68

References to Appendix

1. Bak, B., Christiansen, J. J., and Madsen, E. (1960). *Acta chem. scand.* **14**, 573–576.
2. Beaudet, R. A. (1962). *J. Chem. Phys.* **37**, 2398–2402.
3. Beaudet, R. A. (1963). *J. Chem. Phys.* **38**, 2548–2552.
4. Beaudet, R. A. (1964). *J. Chem. Phys.* **40**, 2705.
5. Beaudet, R. A. and Wilson, E. B. (1962). *J. Chem. Phys.* **37**, 1133.
6. Brown, F. B., Clague, A. D. H., Heitkamp, N. D., Koster, D. F., and Danti, A. (1967). *J. molec. Spectrosc.* **24**, 163–173.

7. Bunker, P. R. and Longuet-Higgins, H. C. (1964). *Proc. Roy. Soc.* (A) **280**, 340–352.
8. Catalano, E. and Pitzer, K. S. (1958). *J. Phys. Chem.* **62**, 873–874.
9. Cignitti, M. and Allen, T. L. (1964). *J. Phys. Chem.* **68**, 1292–1297.
10. Costain, C. C. and Dowling, J. M. (1960). *J. Chem. Phys.* **32**, 158–165.
11. Curl, R. F., Jr. (1959). *J. Chem. Phys.* **30**, 1529–1536.
12. Danti, A. and Wood, J. L. (1959). *J. Chem. Phys.* **30**, 582–584.
13. Dauben, W. G. and Pitzer, K. S. (1956). "Steric Effects in Organic Chemistry", p. 59. Wiley, New York.
14. Dixon, R. N. and Kroto, H. W. (1965). *Proc. R. Soc.* (A) **283**, 423–432.
15. Dixon, W. B. and Wilson, E. B. (1961). *J. Chem. Phys.* **35**, 191.
16. Fateley, W. G. and Miller, F. A. (1962). *Spectrochim Acta.* **18**, 977.
17. Foster, P. D., Rao, V. M., and Curl, R. F. (1965). *J. Chem. Phys.* **43**, 1064.
18. Goodisman, J. (1966). *J. Chem. Phys.* **44**, 2085–2092.
19. Gray, P. and Reeves, L. W. (1960). *J. Chem. Phys.* **32**, 1878–1880.
20. Harris, G. M. and Harris, F. E. (1959). *J. Chem. Phys.* **31**, 1450–1453.
21. Hecht, H. G., Grant, D. H., and Eyring, H. (1960). *Molec. Phys.* **3**, 577.
22. Hirota, E. (1962). *J. Chem. Phys.* **37**, 283.
23. Hirota, E. (1965). *J. Chem. Phys.* **42**, 2071.
24. Karplus, M. and Parr, R. (1963). *J. Chem. Phys.* **38**, 1547–1552.
25. Kasuya, T. (1960). *J. phys. Soc. Japan* **15**, 1273.
26. Kilb, R. W. and Pierce, L. (1957). *J. Chem. Phys.* **27**, 108.
27. Kirtman, B. (1964). *J. Chem. Phys.* **41**, 775–788.
28. Kojima, T. and Nishikawa, T. (1957). *J. Phys. Soc. Japan* **12**, 680.
29. Kojima, T., Breig, E. L., and Lin, C. C. (1961). *J. Chem. Phys.* **35**, 2139.
30. Kopelman, R. (1964). *J. Chem. Phys.* **41**, 1547–1553.
31. Krisher, L. C. (1960a). *J. Chem. Phys.* **33**, 304.
32. Krisher, L. C. (1960b). *J. Chem. Phys.* **33**, 1237–1241.
33. Laurie, V. W. (1959a). *J. Chem. Phys.* **30**, 1210–1214.
34. Laurie, V. W. (1959b). *J. Chem. Phys.* **31**, 1500–1505.
35. Laurie, V. W. (1961). *J. Chem. Phys.* **34**, 1516.
36. Laurie, V. W. and Lide, D. R., Jr. (1959). *J. Chem. Phys.* **31**, 939–943.
37. Li, Y. S., Harmony, M. D., Hayes, D., and Beeson, E. L., Jr. (1967). *J. Chem. Phys.* **47**, 4514–4520.
38. Lide, D. R., Jr. (1958). *J. Chem. Phys.* **29**, 1426–1427.
39. Lide, D. R., Jr. (1959a). *J. Chem. Phys.* **30**, 37–39.
40. Lide, D. R., Jr. (1959b). *Spectrochim. Acta.* **15**, 473–476.
41. Lide, D. R., Jr. (1960). *J. Chem. Phys.* **33**, 1514.
42. Lide, D. R., Jr. and Mann, D. E. (1957). *J. Chem. Phys.* **27**, 868–873.
43. Loos, K. R. and Lord, R. C. (1965). *Spectrochim. Acta.* **21**, 119.
44. Morino, Y. and Hirota, E. (1958). *J. Chem. Phys.* **28**, 185–197.
45. Nelson, R. and Pierce, L. (1965). *J. molec. Spectrosc.* **18**, 344–352.
46. Nishikawa, T. (1957). *J. phys. Soc. Japan.* **12**, 668.
47. Pierce, L. (1956). *Bull. Am. phys. Soc.* **1**, 198.
48. Pierce, L. (1958). *J. Chem. Phys.* **29**, 383.
49. Pierce, L. (1959). *J. Chem. Phys.* **31**, 547–548.
50. Pierce, L. (1961). *J. Chem. Phys.* **34**, 498.
51. Pierce, L. and Hayashi, M. (1961). *J. Chem. Phys.* **35**, 479–485.
52. Pierce, L. and Krisher, L. C. (1959). *J. Chem. Phys.* **31**, 875–882.
53. Pierce, L. and O'Reilly, J. M. (1959). *J. molec. Spectrosc.* **3**, 536–547.

54. Pierce, L. and Peterson, D. H. (1960). *J. Chem. Phys.* **33**, 907–913.
55. Tannenbaum, E., Myers, R. J., and Gwinn, W. D. (1956). *J. Chem. Phys.* **25**, 42.
56. Pitzer, R. M. (1964). *Diss. Abstr.* **25**, 870.
57. Quade, C. R. and Lin, C. C. (1963). *J. Chem. Phys.* **38**, 540–550.
58. Radcliffe, K. and Wood, J. L. (1966). *Trans. Faraday Soc.* **62**, 1678.
59. Risgin, O. and Taylor, R. C. (1959). *Spectrochim. Acta.* **15**, 1036.
60. Schwendeman, R. H. and Jacobs, G. D. (1962). *J. Chem. Phys.* **36**, 1251.
61. Siegel, S. (1957). *J. Chem. Phys.* **27**, 989–990.
62. Stiefvater, O. L. and Sheridan, J. (1963). *Proc. chem. Soc.* 368
63. Stone, R. G., Srivistava, S. C., and Flygare, W. H. (1968). *J. Chem. Phys.* **48**, **1890–1897.**
64. Swalen, J. D. and Costain, C. C. (1959). *J. Chem. Phys.* **31**, 1562–1574.
65. Swalen, J. D. and Stoicheff, B. P. (1958). *J. Chem. Phys.* **28**, 671.
66. Unland, M. L., Weiss, V., and Flygare, W. H. (1965). *J. Chem. Phys.* **42**, 2138–2149.
67. Verdier, P. H. and Wilson, E. B., Jr. (1958). *J. Chem. Phys.* **29**, 340–347.
68. Wang, I. Y. M., Britt, C. O., Cowley, A. H., and Boggs, J. E. (1968). *J. Chem. Phys.* **48**, 812–817.
69. Ward, C. R. and Ward, C. H. (1964). *J. molec. Spectrosc.* **12**, 289.
70. Wyatt, R. E. and Parr, R. G. (1966). *J. Chem. Phys.* **44**, 1529–1545.

Chapter 5

HYDROGEN-BONDED AND CHARGE-TRANSFER SYSTEMS

Since 1960, the direct observation and assignment of low-frequency hydrogen bonding modes has been the subject of intensified research interest. As pointed out by Pimentel and McClellan (1960), these low-lying modes are the most interesting of all the vibrational degrees of freedom of hydrogen-bonded polymers, and until recently they have been the most difficult to observe. The force constants calculated from these frequencies provide information about the hydrogen-bonding potential function as well as being a measure of the bond strength. Also, because these modes are the lowest in frequency, they provide an important contribution to the entropy of the system.

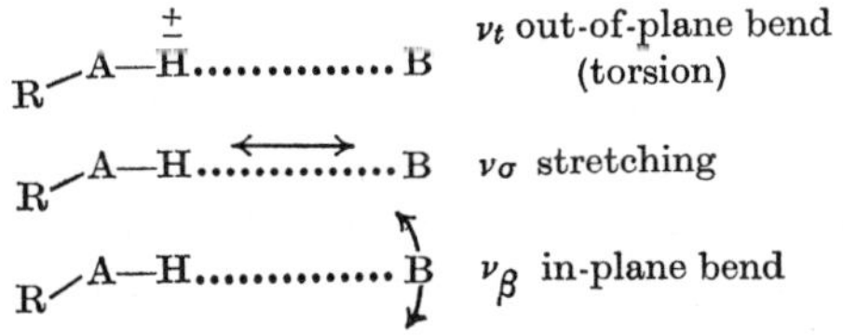

FIG. 1. Low-frequency vibrational modes associated with hydrogen bonding.

The formation of a hydrogen bond may be regarded as a weak intermolecular complex which generates new intermolecular vibrational modes having no counterparts in the separated molecules. Of the six new modes, at least three, ν_σ, the A–B stretch, ν_β the in-plane bend, and ν_t the torsion can often be observed in the far infrared. These are shown in Fig. 1.

I. Assignments of Hydrogen-Bond Modes

As with other vibrational modes occurring in the far infrared, unequivocal assignment of the hydrogen bond modes has met with some difficulty. Recently, however, with advances in instrumentation and several newly developed techniques, definite assignments have been made for several hydrogen-bonded systems.

A. Formic and Acetic Acid

In the far infrared, formic and acetic acid are amongst the most extensively investigated hydrogen-bonded systems. Halford (1946) reported a Raman band near 200 cm^{-1} for liquid formic acid and postulated that this might be the stretching mode of a cyclic dimer. The same work also predicted the hydrogen-bond stretching force constant to be 0.4×10^{-5} dynes cm^{-1}. Bonner and Kirby-Smith (1940) also observed a Raman line in formic acid vapour at 232 cm^{-1}. Millikan and Pitzer (1958) reported two infrared absorptions of formic acid dimer at 237 and 160 cm^{-1}. A normal coordinate analysis by Miyazawa and Pitzer (1959) yielded the hydrogen-bond frequencies of dimeric formic and acetic acids. These calculations could not be confirmed, however, due to a lack of spectral data. Hirota and Nakai (1959, 1960) also measured the far-infrared spectra of formic and acetic acid dimers and some of their deuterated derivatives, but have interpreted their data on the basis of a double minimum in the potential curve for the hydrogen bond (see Section II).

As far as gaseous dimeric formic and acetic acids are concerned, the above mentioned research efforts over the period 1940 to 1960 failed to produce unequivocal assignments for the hydrogen-bonded fundamentals. It has been well established that both formic and acetic acids exist predominantly as hydrogen-bonded dimers in the vapour phase at temperatures below 150°C. The dimers have a planar ring structure of C_{2h} symmetry. An $(R\text{–}COOH)_2$ molecule has twenty-four degrees of vibrational freedom other than the internal vibrations of the R-groups and, for C_{2h} symmetry, eighteen of these can be correlated with the internal vibrations of the monomer units, while the other six can be described as translational and rotational vibrations of the two monomer units against each other.

Since 1960, advances in far-infrared instrumentation have allowed more complete spectral information to be obtained. Carlson *et al.* (1966) were among the first to apply some of the newer far-infrared instrumentation to this problem. Their results for gas phase, dimeric formic and acetic acid are given in Table I and Fig. 2. Only three of the six funda-

TABLE I. Low-frequency Infrared-active Vibrations of Formic and Acetic Acid Dimers.

Species	No.	Approximate description	Assignment (cm^{-1}) $(HCOOH)_2$ Carlson *et al.*	$(HCOOH)_2$ Miyazawa and Pitzer	Carlson *et al.* $(HCOOD)_2$	$(CH_3COOH)_2$ Carlson *et al.*	$(CH_3COOH)_2$ Miyazawa and Pitzer	Carlson *et al.* $(CD_3COOH)_2$
b_u	ν_3	H-bond stretch	248	248	240	190	180	180
a_u	ν_4	Ring torsion (around the line connecting the C atoms)	164	160	158	168	80	157
	ν_5	Out-of-plane O–H···O bending	68	60	68	50	46	50

mentals associated with stretching and bending of hydrogen bonds should be infrared active. For formic acid, reasonable agreement is observed between these frequencies and those calculated by Miyazawa

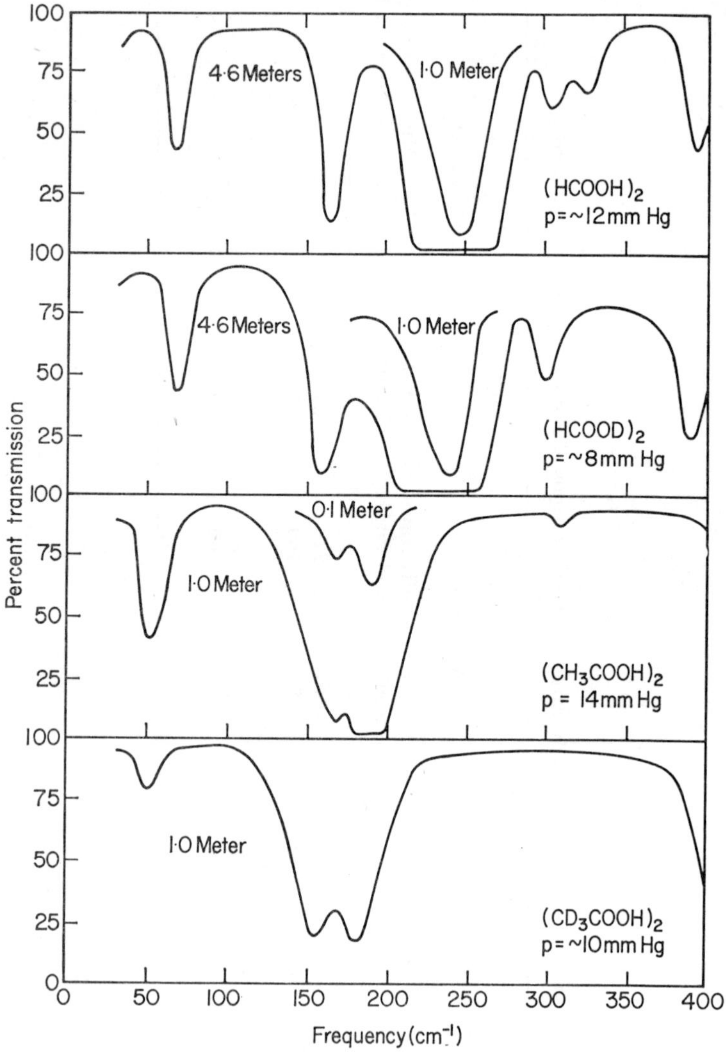

FIG. 2. Infrared spectra of dimeric formic and acetic acid vapours at room temperature. Sample pressures and path lengths as indicated on spectra.

and Pitzer (1959). Low-frequency infrared-active hydrogen bond modes of dimeric formic acid were presumed to be at 248 (b_u), 164 (a_u) and 68 (a_u) cm^{-1}. However, the frequencies presented by Carlson and co-workers for acetic acid do not agree with the calculations of Miyazawa and Pitzer, especially for ν_4 which has been calculated to appear at 80 cm^{-1}, not corresponding to the 168 cm^{-1} band which was observed. Carlson and co-workers suggested that the calculated value of 80 cm^{-1} might be in error, but also suggested that the 168 cm^{-1} band may be the result of Fermi resonance between the combination tone of 178 cm^{-1} [128 (ν_6, b_g) + 50 (ν_5, a_u) and the 180 cm^{-1} (ν_3, b_u)], both the 180 cm^{-1} and 128 cm^{-1} frequencies being calculated values of Miyazawa and Pitzer.

Jakobsen *et al.* (1967), using normal coordinate analysis and far-

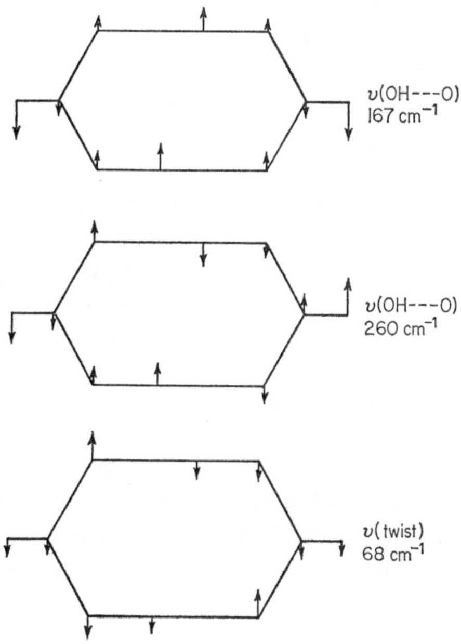

FIG. 3. Cartesian displacements of out-of-plane hydrogen bond modes of formic acid.

infrared spectra, finally established the assignment of both formic and acetic acids. Their calculations revealed that the 68 cm^{-1} band in formic acid vapour was actually the twisting mode, while the 167 cm^{-1} absorption was the bending mode, contrary to the assignment of Carlson and co-workers. A description of these modes is given in Fig. 3 and Table II.

For acetic acid, Jakobsen and co-workers (1967) showed that the 168 cm^{-1} absorption cannot be the bending mode (an explanation favoured by Carlson and co-workers) and concluded that Fermi resonance as described previously was the explanation for this frequency.

Assignments given in the above paper have been confirmed in a

TABLE II. Hydrogen Bond Frequencies of the Gaseous Cyclic Dimers of Formic and Acetic Acid[a].

$(HCOOH)_2$		$(CH_3COOH)_2$		
Calculated	Observed	Calculated	Observed	cm^{-1}
260	—	210	—	
249	248	187	188	
224	—	—	168	
167	173	130	—	
91	—	81	—	
68	68	79	—	
—	—	54	50	

[a] Jakobsen *et al.* (1967).

further important contribution relating to crystalline formic acid (Mikawa *et al.* 1967). Infrared spectra to 70 cm^{-1} for the β-polymorph of $(HCOOH)_n$, $(DCOOH)_n$ and $(HCOOD)_n$ were obtained and assigned on the basis of an infinite hydrogen-bonded chain having a factor group isomorphous with C_{2v}. Using a simple Urey-Bradley force field for the in-plane vibrations of the system, a force constant of 0.24 md $Å^{-1}$ for the stretching of the O···H bond was obtained. The normal coordinate calculation accounted satisfactorily for all of the isotope shifts.

Very recently, low-frequency Raman data on HCOOH and HCOOD were recorded on liquid and solid samples, and, by analogy with the normal modes in the dimer, bands were assigned to the normal modes of a polymeric chain with two molecules in each chain unit cell (Blumenfeld and Fast, 1968).

B. Cresols and Chlorophenols

Hurley *et al.* (1966) established the following three criteria for making more positive assignments of hydrogen-bond modes in their study of cresols and chlorophenols; these criteria should be generally applicable to other hydrogen bonded systems. Firstly, the bands associated with hydrogen bond stretches or bends should disappear completely when a group incapable of hydrogen bonding is substituted for the hydrogen atom (for example $-OCH_3$ substituted for $-OH$). Secondly, the hydrogen bond frequencies should be weak in the Raman spectrum due to small changes in polarizability; although these modes may be allowed by the selection rules, they will have low intensity. Finally, the far-infrared absorptions due to hydrogen bonding should be broad and asymmetric. Although this latter criterion is not unique to hydrogen-bond absorptions, it is a general feature in both the mid and far infrared. The mid-infrared A–H stretch is usually extremely broad and often asymmetric, due, in part, to the contributions of polymers with varying degrees of association and with different band centres and intensities. This will also be the case for ν_σ and ν_β in pure liquids, and, in addition, the anharmonicity of the vibration is often apparent in low-frequency absorptions.

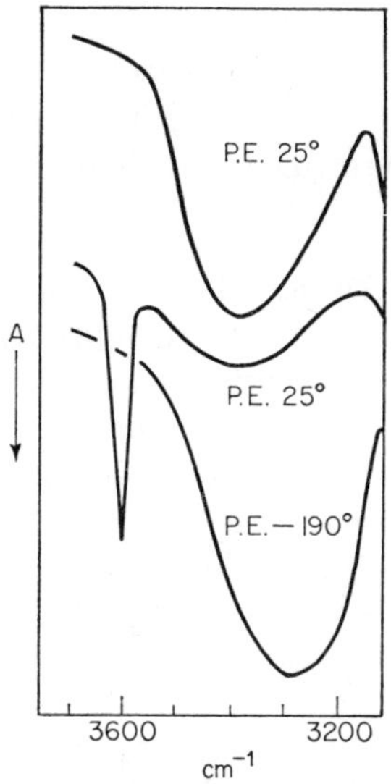

FIG. 4. The hydrogen bond stretching frequency of *p*-cresol in polyethylene matrix.

Using all of the above criteria, Hurley and co-workers were able to assign ν_σ successfully for various cresols and chlorophenols. For example, ν_σ in *o*-, *m*-, and *p*-cresol is assigned at 124, 146 and 124 cm^{-1}, respectively.

The above frequencies are in excellent agreement with those of Jakobsen and Brasch (1965) who assigned ν_σ in the same series to 126, 148 and 130 cm^{-1}, using the polyethylene matrix method. This technique, originally reported by Brasch and Jakobsen (1964), involves the use of polyethylene as a solvent in dilution studies. It has been shown that dilution of hydrogen-bonded systems is one of the most convenient methods of establishing the assignments of hydrogen-bond vibrations, although in the far infrared very few solvents have been found suitable for such studies. Polyethylene, however, being extremely well-suited for window material in this region, was found to be equally usable as a solvent. Jakobsen and Brasch (1965) have observed that a hydrogen-bonded system is initially absorbed into the polyethylene matrix as the polymer, but by suitable reheating of the polyethylene matrix the compound can gradually be converted to a monomer or "free" hydrogen-bonded form. This is illustrated in Fig. 4 which shows the –OH stretching of *p*-cresol in polyethylene matrix; all three of the spectra are of the same polyethylene sample. The upper curve is the spectrum after incorporation of the sample in the film. All the –OH appears to be in the bonded form (broad, diffuse absorption). The centre spectrum was obtained after gently reheating the polyethylene film several times, most of the –OH being now in the monomeric state, as evidenced by the rather sharp absorption at approximately 3600 cm^{-1}. The lower curve is the same sample cooled to liquid nitrogen temperature. Obviously, the reformation of polymeric –OH is evident. The three main advantages of this technique are:

1. By simple heating of the polyethylene film, the "bonded" –OH can be converted to "free" –OH and the disappearance of –OH···O– vibrations can be followed in the far infrared.
2. When the compound is in the monomeric –OH state, cooling apparently converts it back to a hydrogen-bonded state and the assignment of the –O–H···O– vibrations can be checked by following the appearance of these vibrations. In effect, dilution studies are possible with temperature rather than concentration as the variable.
3. Low-temperature spectra can be obtained with the interferences of lattice frequencies minimized. These low-temperature spectra are desirable, since the –O–H···O– vibrations are often broad, weak bands and cooling tends to sharpen the bands, thus making them easier to locate.

C. Other Hydrogen-Bonded Systems

Several other hydrogen-bonded systems have been studied in the far infrared, assignments usually being made on the basis of dilution studies. Lake and Thompson (1966) have studied a homologous series of aliphatic alcohols and have assigned ν_σ in the region 180–110 cm^{-1}. The intensities of these bands were directly proportional to the concentration of alcohol in benzene. The observed frequencies were in good agreement with those assigned for carboxylic acid derivatives by Jakobsen *et al.* (1967) and Stanevich (1963) as well as those for liquid and solid phenols observed by Jakobsen and Brasch (1965). This same work also reported the assignment of the torsional mode, ν_t, from vapour state spectra. The band thus assigned was not observed in the liquid and was assumed to be due to monomeric forms. The frequencies fell in the region 197 cm^{-1}–217 cm^{-1}.

Brasch *et al.* (1968) have very recently established that, at least for the phenol system, the ν_σ frequency is definitely dependent on the strength of the hydrogen bond. This was shown by isolating two polymorphic forms of phenol and measuring both mid- and far-infrared spectral regions. Far-infrared spectra were obtained with an interferometer, using a diamond-window, high-pressure cell. The relevant results are shown in Table III.

TABLE III

	ν (O–H)	ν_σ	Other bands	
Polymorph 1	3200, *vs*, broad	175, *s*	80, *mw*, 47, *w*	cm^{-1}
Polymorph 2	3400, *s*, sharp	135, *m*	92, *m*, 49, *w*.	

Since ν(OH) is generally accepted as a measure of hydrogen bond strength (Pimentel and McClellan, 1960), and since ν_σ correlates with ν(OH) it was argued that it is therefore a function of hydrogen-bond strength. However, it was pointed out that, due to mass effects, this correlation may not hold for all hydrogen-bonded systems.

Ginn and Wood (1967) have studied hydrogen-bonded complexes of phenol with trimethylamine, triethylamine and pyridine. Solutions of phenol in each of these three materials showed a medium or strong band below 150 cm^{-1} which was absent in the unmixed components; and this absorption has been assigned to ν_σ. Concentrated solutions of phenol in carbon tetrachloride also exhibits a band in this region which results from hydrogen bonding.

A study of the effect of environment on the intermolecular stretching frequency, ν_σ, for the phenol-pyridine complex has been made by Hall and Wood (1968). The mole fraction of phenol was held constant at half the value of pyridine throughout and the mole fraction of CCl_4 varied from 0.85 to zero. The effect of the variation on the ν_σ band at ~134 cm^{-1} was that it increased in frequency as the proportion of CCl_4 was reduced; that is, as the environment became more polar. This is the same direction as the frequency shifts with solvent observed for the intermolecular stretching modes of charge transfer complexes and the implications of this are discussed. The effect on ν_s could not be observed due to the breadth and complex profile of the absorption.

II. The Double-Minimum Potential Function for Hydrogen Bonds

Periodically, over the past several years, various research work has proposed and offered evidence for the existence of a double-minimum potential function for the hydrogen bond (Fig. 5). An excellent review by Cannon (1958) cites various authors who propose such a potential function based on several sources of physical evidence. None of this data, however, was obtained from measurements in the far infrared. Several workers have attempted to explain the broadness (and splitting) of the –OH stretching frequency as due to proton tunnelling across the potential barrier. The most direct evidence should be observable in the far infrared where absorptions due to transition between the split levels ($v_0+ \rightarrow v_0-$ and $v_1+ \rightarrow v_1-$) may occur. Such evidence has been relatively scarce and much of that which has been proposed has subsequently been disproven. Hirota and Nakai (1959) suggested that shoulders

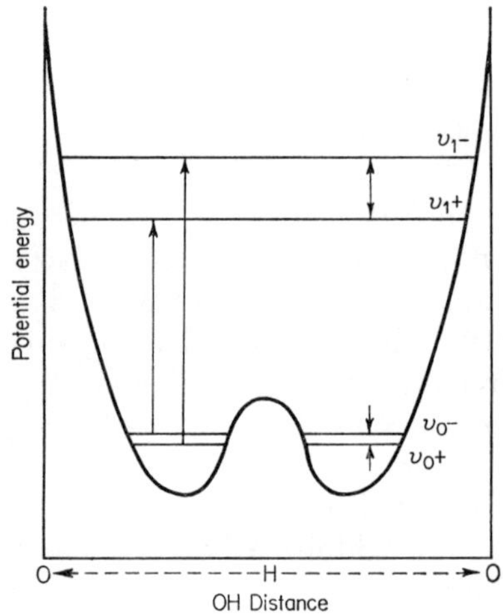

FIG. 5. The double minimum potential function for the hydrogen bond.

appearing at 270 and 175 cm^{-1} in the far-infrared spectrum of formic acid were due to the motion of the hydrogen-bonded proton in the double minimum potential. It has been shown that this interpretation was in error and the far-infrared spectrum of formic acid has since been adequately assigned by Jakobsen *et al.* (1967) and others.

Hadzi (1961) has reported the observation of the $v_0+ \rightarrow v_0-$ transition in several solid ferroelectric crystals in the range 121–151 cm^{-1}. These absorptions were not observed in the deuterated materials and it was suggested that they probably shifted to below 100 cm^{-1}. This apparent deuteration shift supported the above assignment rather than the alternative suggestion that these absorptions may simply be lattice modes.

Support for the existence of a double potential minimum for deuterated chromous acid ($DCrO_2$) was given by Snyder and Ibers (1962). These authors observed markedly different spectra for $HCrO_2$ and $DCrO_2$ over the region 4000–300 cm^{-1}, differences which could not be explained simply on the basis of the mass change going from hydrogen to deuterium. For example, the O–H stretching band is observed at 1640 cm^{-1} in $HCrO_2$ while in $DCrO_2$, allowing for mass change and possible anharmonicity, this band would be expected in the region 1200–1300 cm^{-1}. No bands were observed in this region, and instead strong bands were observed near 1750 cm^{-1}. In addition, the O–H stretching band in $HCrO_2$ is a singlet, while in $DCrO_2$, the O–D stretch is a doublet. The ratio of the frequencies of the hydrogen-bonded bending modes is 1.44 at 24°C and 1.49 at −196°C. The harmonic oscillator value is 1.386 and in other strongly hydrogen-bonded systems has been found to be 1.34–1.38. Finally, two lattice bands were observed for $HCrO_2$ while six were found for $DCrO_2$.

These observations were explained on the basis of a difference in the potential functions of $HCrO_2$ and $DCrO_2$. If the shapes of the potential functions are different, then the transitions may not be between the same levels and the band representing the stretching mode would be in a position not predicted by the harmonic value. A high barrier in the O–D–O potential could result in two sets of closely-spaced levels, and the allowed transitions between these levels would give rise to the doublet in the O–D stretching region. The observed ratio, ν_{OHO}/ν_{ODO}, is larger than that predicted by mass effects alone. This is explained by the fact that the "hydrogen" bond in $DCrO_2$ may be weaker than that in $HCrO_2$, as the frequencies usually increase with hydrogen-bond strength. Lastly, as the authors point out, if the barrier in $HCrO_2$ is low or perhaps zero, the hydrogen atom is symmetrically located and the selection rules for D_{3d} symmetry are valid. On the other hand, the deuterium in $DCrO_2$ tunnels from one side of the barrier to the other at a frequency low compared with the observed frequency of the bands. Hence, the deuterium atom is in effect not at the centre of symmetry, and all nine fundamentals are active. Thus, instead of two lattice modes as predicted and observed for $HCrO_2$, all six lattice modes of $DCrO_2$ were observed.

Similar evidence has been offered by Benoit (1963) for solid $HCoO_2$ and $DCoO_2$, and he also suggests a difference in the potential function between the –O–H and –O–D bonds.

These experimental observations lend strong support for the existence of a double minimum potential function, at least in the –O–D–O bond of the hydrogen-bonded crystalline materials studied. It should be noted, however, that recent crystallographic studies (Christensen 1966) have given evidence for a new polymorphic form of $HCrO_2$, a fact which could reasonably account for the differences observed in the infrared spectra. In addition, the above cited evidence prompted Ginn and Wood (1965) to re-examine ν_σ, the H– or D– bond stretching vibration, in phenol-amine complexes (Ginn and Wood 1964) for possible evidence for a double minimum potential. They concluded that a shoulder which was observed in the phenol-pyridine complex at 143 cm^{-1} and shifted to 139 cm^{-1} on deuteration was not the result of proton tunnelling and that their results were consistent with no change in the potential energy function on deuteration.

It is apparent from the above discussion that the question of the double minimum potential function for the hydrogen bond is far from answered and much work is still to be done.

III. Charge Transfer Complexes

Far-infrared spectroscopy has proved to be a very useful method of investigating charge transfer complexes and there has been considerable activity in the last two to three years. Much of the work relates to complexes of various donors with halogens and mixed halogens. Yada *et al.* (1962) assigned far-infrared bands at 185 cm^{-1} and 145 cm^{-1} in the $(CH_3)_3NI_2$ complex to I–I stretching and N–I stretching, respectively. More recently Yokobayashi *et al.* (1968) reported far-infrared spectra of the trimethylamine complexes of I_2, IBr and ICl. Their assignments are summarized in Table IV.

TABLE IV

$(CH_3)_3NI_2$	$(CH_3)_3N.IBr$	$(CH_3)_3N.ICl$		
172	206	249	I–X str.	cm^{-1}
146	172	196	N–I str.	
73	90	147	N–I–X def.	

The majority of the work relates to complexes of pyridine and substituted pyridines with halogens and mixed halogens. If these systems are examined in a polar medium, bands due to trihalide ions of the type X_3^- and $X'X_2^-$ have been observed and this has led to an interest in their low frequency spectra. Complexes of pyridine (py) with I_2, IBr and ICl were investigated by Ginn and Wood (1966). In the py-I_2 system a strong band at 167 cm^{-1} was observed, together with a weaker absorption at 137 cm^{-1}. By variation of the proportions of pyridine,

I_2 and *n*-hexane, it was found that as the proportion of *n*-hexane increased the 167 cm^{-1} band remains strong but moves to 180 cm^{-1}, whereas the lower band weakens continuously, suggesting that it might be due to an ionic species, that is I_3^-. A weak band at ~120 cm^{-1} was tentatively assigned to the intermolecular N–I stretch and the 167 cm^{-1} band to the I–I stretch. Some of the assignments are summarized in Table V.

TABLE V

	ν (cm^{-1})	species	mode
I_2 in excess pyridine	167	py I_2	I–I str.
	137	I_3^-	a.s. str.
	120	py I_2	N–I str.
py IBr in pyridine	195	py IBr	I–Br str.
	174	IBr_2^-	a.s. str.
	144	py IBr	N–I str.
py ICl in pyridine	277	py ICl	I–Cl str.
	223	ICl_2^-	a.s. str.
	160	py ICl	N–I str.

A simple three-centre molecular orbital treatment was found to account satisfactorily for the observations on these complexes. It was suggested that equilibria of the following type account for the presence of the polyhalide ions:

$$2\ \text{py IX} \rightleftarrows \text{py}_2\text{I}^+ + \text{IX}_2^-.$$

More recently, Ginn *et al.* (1968) have studied the $pyBr_2$ and pyBrCl systems using far infrared and laser Raman spectra. The spectra were consistent with C_{2v} symmetry with the Br_2 lying along the C_2 axis. The assigned X–X′ and N–X stretching frequencies are summarized in Table VI.

TABLE VI

	ν (cm^{-1})	mode
$pyBr_2$ in C_6H_6	229	Br–Br str.
	128	N–Br str.
pyBrCl in C_6H_6	307	Br–Cl str.
	144	N–Br str.

In the Raman spectrum of the $pyBr_2$ system in benzene, two strong bands at 163 cm^{-1} (assigned to Br_3^-) and at 215 cm^{-1} (assigned to Br–Br stretch in the complex) were observed. This disagrees with some earlier Raman work by Klaboe (1967) who found a weak band at 302 cm^{-1} (assigned to the benzene-Br_2 complex) and a stronger band at 281 cm^{-1} (assigned to the pyridine-Br_2 complex).

In the solid state it was shown (Ginn *et al.* 1968) that the complex shows considerable differences from the solution, and all the low frequency bands could be accounted for by the py_2Br^+ and Br_3^- ions. The complexes were predominantly un-ionized in non-polar solvents, but in polar media similar equilibria to those occurring in the pyI_2, pyIBr and pyICl complexes were indicated by the additional bands. The effects of solvent on each complex were examined, and as the polarity of the environment increases, the N–Br stretching frequency rises and the Br–X stretching frequency falls, both these changes being in the same direction as those produced by increasing electron transfer. Further studies on the effect of solvent on the I_2, IBr, ICl and ICN complexes of pyridine (Haque and Wood 1967a) and γ-picoline (Haque and Wood 1967b) have also been made.

Complexes of I_2 with substituted pyridines in cyclohexane solutions were studied interferometrically by Lake and Thompson (1967). Bands were observed in each of ten complexes in the region of 180 cm^{-1} and between 65–95 cm^{-1}, these being assigned to I–I stretching and pyridine–I stretching respectively. Measurements of the intensities were made and equilibrium constants for the formation of the complexes deduced. Force constant calculations (using a simple VFF) were also made and the results discussed in terms of mass, steric and electronic effects. Very

TABLE VII. Stretching Frequencies and Force Constants for Some Complexes of Br_2 with Substituted Pyridines.

Donor	ν_1 (Br–Br str.) (cm^{-1})	ν_2 (N–Br str.) (cm^{-1})	$k_{12} = 0$		$k_{12} = 0.3$	
			k_1	k_2	k_1	k_2
Pyridine	231	120	1.07	0.53	1.44	0.46
2 Methyl pyridine	226	113	1.02	0.52	1.39	0.50
3 Methyl pyridine	228	119	1.00	0.60	1.40	0.49
4 Methyl pyridine	225	115	1.00	0.55	1.37	0.46
2,4 Dimethyl pyridine	223	110	0.98	0.55	1.35	0.46
2,6 Dimethyl pyridine	230	100	1.11	0.42	1.44	0.39
3 Chloro pyridine	243	98	1.27	0.41	1.58	0.39
3 Bromo pyridine	242	89	1.26	0.41	1.56	0.38

recently the same workers (*vide supra* 1968) have made a similar study on complexes of Br_2 with substituted pyridines. Bands in the region 90–120 cm^{-1} were assigned to the intermolecular py–Br stretch and in 220–250 cm^{-1} region to stretching of Br–Br. A band at 187 cm^{-1} associated with the Br_3^- ion was also observed.

From the two frequencies and making the approximation of a three-body system, two stretching force constants k_1(X–X) and k_2(N–X) were computed for the series. When a stretch-stretch interaction term k_{12} was included (this being set equal to 0.3 md/Å) the order of force constants remained the same for the various donors. Some of the results are summarized in Table VII.

The intermolecular stretching frequency is significantly affected by the solvent and a few of the values obtained for ν_1 and ν_2 of 4-methyl pyridine-I_2 complexes are listed below (Table VIII).

TABLE VIII. Variation of ν_1 and ν_2 (cm^{-1}) with Solvent for 4-Methyl Pyridine –I_2 Complexes.

Solvent	ν_1 (cm^{-1})	ν_2 (cm^{-1})
Cyclohexane	181	88
n-Pentane	182	85
Carbon tetrachloride	178	94
Benzene	171	100
o-Xylene	172	100

Yagi *et al.* (1967) have examined the influence of complex formation on the IBr stretching frequency in complexes with various donors. Both the frequency and intensity were strongly influenced by the strength of the IBr donor interaction. The frequency varied from 261 cm^{-1} for IBr in CCl_4 to 205 cm^{-1} for pyIBr in C_6H_6. Spectra of some solid state complexes of ICl and IBr with pyridine, 4,4′ bipyridine and 2,2′ bipyridine were also reported and force constant calculations were made for these and the other systems.

It was mentioned earlier that equilibria of the type

$$2\ \text{py XX}' \rightleftharpoons [\text{py}_2\text{X}]^+ + [\text{XX}'_2]^-$$

complicate the assignment of bands in the pyXX′ complexes and bands due to these ions have been assigned. The polyhalide ions of types X_3^- and XX_2' have been investigated in some detail by several authors and are discussed later. The iodopyridinium ion, py_2I^+, was investigated by Creighton *et al.* (1966) who examined the far-infrared spectra of $[py_2I]^+BF_4^-$ and $[py_2I]^+ClO_4^-$ in pyridine solution, together with the

Raman spectrum of $[py_2I]^+BF_4^-$ in pyridine. Bands were observed at 172 cm^{-1} in the infrared and 181 cm^{-1} (pol) in the Raman, and were thus assigned to antisymmetric and symmetric stretching respectively of a linear N–I–N skeleton.

Iodine complexes of 1,4-dithian and 1,4-diselenan were measured by Hendra and Sadasivan (1965). In the dithian-I_2 complex the bands at 158 cm^{-1} and 212 cm^{-1} were assigned to I–I stretching and S–I stretching respectively. The corresponding modes in the diselenan-I_2 complex were assigned to bands at 135 cm^{-1} and 170 cm^{-1}.

Complexes of halogens and mixed halogens with aromatic molecules have attracted a good deal of attention (see Person *et al.* 1967 and references therein) including vibrational investigations. Infrared spectra of the solid benzene-Br_2 complex (Person *et al.* 1967) showed definitely that neither the benzene nor the bromine molecules were located on a site possessing a centre of symmetry, and from analysis of the complexed benzene a site symmetry of C_{3v} was deduced. This is in apparent contradiction to the results of a previous X-ray study (at –50°C) in which it was reported that the crystals were monoclinic. However, the structure is very close to one with a rhombohedral unit cell with C_6H_6 and Br_2 located on sites of C_{3v} symmetry and it was suggested that as the infrared measurements were carried out at –200°C, a different crystal form may be involved and that only a relatively minor adjustment of the crystal structure is needed to pass from one form to the other. Intensity changes resulting from complex formation are also discussed.

The occurrence of trihalide ions of the types X_3^- and XY_2^- when the pyridine-halogen complexes are examined in a polar medium was mentioned above, and a brief note of some of the low-frequency results for these is appropriate.

The first important paper on the vibrational spectra of systems of this type was given by Person *et al.* (1961) who measured infrared and Raman spectra of ICl_2^-, ICl_4^-, $BrCl_2^-$ and Br_3^- and assigned the two stretching frequencies in each species. From these results the XY stretching force constant and the interaction stretching force constants between the two XY bonds were calculated. These were interesting in that the stretching force constants were about one half of the value in the free halogens and the interaction constants were relatively very large, being about 35% of the value of the stretching force constant. These are summarized in Table IX.

Hayward and Hendra (1967) examined the far infrared and Raman spectra of IBr_2^- and I_3^- and a very thorough study of ICl_2^-, IBr_2^-, I_3^-, I_2Br^- and $BrICl^-$ has been reported by Maki and Forneris (1967). In the latter work stretching and bending frequencies were assigned and

TABLE IX. Bond Stretching Force Constants (f_r) and Interaction Stretching Constants (f_{rr}) for ICl_2^-, $BrCl_2^-$ and Br_3^- Ions and Comparison with Appropriate Free Halogen.

Species	f_r (md/Å)	f_{rr} (md/Å)
ICl_2^-	1.00	0.36
ICl	2.38	—
$BrCl_2^-$	1.29	0.26
BrCl	2.83	—
Br_3^-	0.91	0.32
Br_2	2.45	—

force constants computed for all the species. As above (Person *et al.* 1961), the stretching force constants were found to be about half the value in the corresponding halogen and interhalogen molecules. Some of the assignments are summarized in Table X.

TABLE X.

Cation	Anion	ν_1	ν_2	ν_3 (cm^{-1})
Cs^+	ICl_2^-	268	129	218
$(CH_3)_4N^+$	IBr_2^-	160	98	171
Cs^+	I_3^-	103	69	149
Cs^+	BrI_2^-	168	84	117

The "BrICl$^-$" ion showed a surprisingly complex spectra and it was suggested that a mixed crystal of ICl_2^- and IBr_2^- is present.

X-ray measurements on some trihalide ions have shown that they are often slightly asymmetric (for example the two bonds are unequal in length or there is a slight departure from linearity) and, for a given ion, this may vary with the cation. This is in agreement with the changes observed in the infrared and Raman spectra on passing from one compound to another with the same polyhalide ion. The differences can be rationalized by assuming certain differences in the crystal structure, and thus from the vibrational data predictions about the symmetry of the trihalide ions in different compounds can be made.

References

Benoit, A. (1963). *Spectrochim. Acta.* **19,** 2011.

Blumenfeld, S. M. and Fast, H. (1968). *Spectrochim. Acta.* **24A,** 1449.

Bonner, L. and Kirby-Smith, J. S. (1940). *Phys. Rev.* **57,** 1078.

Brasch, J. W. and Jakobsen, R. J. (1964). *Spectrochim. Acta.* **20**, 1644.
Brasch, J. W., Jacobsen, R. J., Fateley, W. G., and McDevitt, N. T. (1968). *Spectrochim. Acta* **24A, 203.**
Cannon, C. G. (1958). *Spectrochim. Acta* **10**, 341.
Carlson, G. L., Witkowski, R. E., and Fateley, W. G. (1966). *Spectrochim. Acta.* **22**, 1117.
Christensen, A. N. (1966). *Inorg. Chem.* **5**, 1452.
Creighton, J. A., Haque, I., and Wood, J. L. (1966). *Chem. Commun.* 229.
Ginn, S. G. W. and Wood, J. L. (1964). *Proc. chem. Soc.* 370.
Ginn, S. G. W. and Wood, J. L. (1965). *Chem. Commun.* 628.
Ginn, S. G. W. and Wood, J. L. (1966). *Trans. Faraday Soc.* **62**, 777.
Ginn, S. G. W. and Wood, J. L. (1967). *Spectrochim. Acta* **23A**, 611.
Ginn, S. G. W., Haque, I., and Wood, J. L. (1968). *Spectrochim. Acta* **24A**, 1531.
Hadzi, D. (1961). *Spectrochim. Acta* **17**, 1445.
Halford, J. O. (1946). *J. Chem. Phys.* **14**, 395.
Hall, A. and Wood, J. L. (1968). *Spectrochim. Acta* **24A**, 1109.
Haque, I. and Wood, J. L. (1967a). *Spectrochim. Acta* **23A**, 959.
Haque, I. and Wood, J. L. (1967b). *Spectrochim. Acta* **23A**, 2523.
Hayward, G. C. and Hendra, P. J. (1967). *Spectrochim. Acta* **23A**, 1937.
Hendra, P. J. and Sadasivan, N. (1965). *Spectrochim. Acta* **21**, 1127.
Hirota, K. and Nakai, Y. (1959). *Bull. chem. Soc. Japan* **32**, 769.
Hirota, K. and Nakai, Y. (1960). *J. chem. Soc. Japan* **81**, 881.
Hurley, W. J., Kuntz, Jr., I. D., and Leroi, G. E. (1966). *J. Am. chem. Soc.* **88**, 3199.
Jakobsen, R. J. and Brasch, J. W. (1965). *Spectrochim. Acta.* **21**, 1753.
Jakobsen, R. J., Mikawa, Y., and Brasch, J. W. (1967). *Spectrochim. Acta.* **23A**, 2199.
Klaboe, P. (1967). *J. Am. chem. Soc.* **89**, 3667.
Lake, R. F. and Thompson, H. W. (1966). *Proc. R. Soc.* A, **291**, 469.
Lake, R. F. and Thompson, H. W. (1967). *Proc. R. Soc.* A, **297**, 440.
Lake, R. F. and Thompson, H. W. (1968). *Spectrochim. Acta* **24A**, 1321.
Maki, A. G. and Forneris, R. (1967). *Spectrochim. Acta* **23A**, 867.
Mikawa, Y., Brasch, J. W., and Jakobsen, R. J. (1967). *J. molec. Spectrosc.* **24**, 314.
Millikan, R. C. and Pitzer, K. S. (1958). *J. Am. chem. Soc.* **80**, 3515.
Miyazawa, T. and Pitzer, K. S. (1959). *J. Am. chem. Soc.* **81**, 74.
Person, W. B., Anderson, G. R., Fordemwalt, J. N., Stammreich, H., and Forneris, R. (1961). *J. Chem. Phys.* **35**, 908.
Person, W. B., Cook, C. F., and Friedrich, H. B. (1967). *J. Chem. Phys.* **46**, 2521.
Pimentel, G. C. and McClellan, A. L. (1960). "The Hydrogen Bond". Freeman, San Francisco, California.
Snyder, R. G. and Ibers, J. A. (1962). *J. Chem. Phys.* **36**, 1356.
Stanevich, A. E. (1963). *Optics Spectrosc.* **15**, 243.
Yada, H., Tanaka, J., and Nagakura, S. (1962). *J. molec. Spectrosc.* **9**, 461.
Yagi, Y., Popov, A. I., and Person, W. B. (1967). *J. Phys. Chem.* **71**, 2439.
Yokobayashi, K., Watari, F., and Aida, K. (1968). *Spectrochim. Acta.* **24A**, 1651.

Chapter 6

COMPLEX INORGANIC SYSTEMS

I. Introduction

A. General

As indicated in Chapter 1, the heavy atoms and low force constants associated with many inorganic systems lead to low vibrational frequencies, and over the past few years a tremendous volume of data has been accumulated, much of it relating to metal-ligand stretching modes. This has largely resulted from the availability of commercial far-infrared spectrometers and interferometers which have extended the frequency range down to 10 cm^{-1}, together with laser-excited Raman spectrometers which enable the spectra of coloured substances and solids to be obtained with much greater ease than before.

Infrared spectroscopy has been used extensively in the study of inorganic chemistry previous to the advent of commercial low-frequency spectrometers. However, such measurements were mostly concerned with internal vibrations of the ligand and only more rarely with vibrations mainly associated with the metal-ligand skeleton of the molecule.

More recently much effort has been concentrated on measurements of metal-ligand skeletal vibrations, both in the infrared and Raman, with a consequent interest in the low-frequency modes of polyatomic ligands. The two chief objectives here are (a) inference of structure from the pattern of infrared and Raman frequencies (see Section V) and (b) deductions about the nature of the metal-ligand bond from frequencies and derived force constants. Low-frequency spectra of systems in which metal-metal bonds occur (including the so called "metal-cluster" compounds) are also currently exciting interest. As many systems may only be studied in the solid state (often due to lack of solubility in suitable solvents), difficulties in assignment of fundamentals arise from the presence of bands due to lattice vibrations at these low frequencies. Hence, there is a good deal of interest in the lattice modes of inorganic systems. Application of high pressure techniques (see Chapter 3) to distinguish between internal modes and lattice modes is currently under investigation.

The most widely investigated and probably the best understood aspect of the work on metal-ligand vibrations is that on metal-chlorine systems and, to a lesser extent, metal-bromine and metal-iodine systems. Most of these frequencies lie below 400 cm^{-1}. Apart from the simple halide ions the commonest ligands are those in which the donor atom is carbon, nitrogen or oxygen. The vibrational frequencies associated with M–O, M–C and M–N bonds cover large ranges which extend into the low-frequency range, and a very large volume of data now exists. An exhaustive survey of the subject of metal-ligand vibrations has been recently given by Adams (1967) covering the literature up to the end of 1965.

Ligands involving non first-row element donor atoms (such as P, As, Sb, S, Se and Te, and so on) usually require low-frequency measurements for location of the modes associated with the metal-ligand skeleton. Also, the presence of the heavier atoms in the ligand gives rise to more low-frequency ligand modes which can confuse the interpretation and increase the likelihood of coupling between internal ligand and metal ligand vibrations. With the accumulation of a large amount of data, various correlations of stretching frequencies (and force constants), with, for example, stereochemistry, oxidation state, coordination number, strength of metal-ligand bonds, and so on, have been made.

It will be impossible in this chapter to give an exhaustive literature coverage, but an attempt will be made to illustrate the more important trends and correlations and indicate the potential and limitations of far-infrared spectroscopy in some of its applications to inorganic chemistry, including some complex oxide systems. Some cautionary points will also be made, as false conclusions are sometimes drawn through uncritical use of the experimental data.

In a discussion of inorganic systems it is rather artificial to isolate infrared from Raman data, and, in fact, the Raman method has a number of inherent advantages: (1) in systems of high symmetry (for example, octahedral MX_6 and tetrahedral MX_4 species) there are more frequencies active in the Raman than in the infrared (see Table XXXI–XXXIV), (2) the high polarizabilities associated with the heavier atoms of inorganic systems lead to intense Raman lines, and (3) the relative ease with which polarization data are obtained on liquid and solution phase samples. Ideally, the two techniques are used complementarily; hence, in this chapter, Raman data will be considered in conjunction with far-infrared data where relevant.

B. Group Frequencies

The concept of group vibrations (that is, normal modes in which the potential energy of the mode is essentially localized in one bond or group of bonds within the molecule) is well known in organic chemistry where the force constants and masses involved have limited ranges, whereas in inorganic chemistry there is a much greater variety. The possibility of mixing together of bond vibrations (provided they are of the same symmetry) in a normal mode increases where bonds connecting atoms of similar masses have similar force constants. When this occurs, the potential energy distribution in the normal mode may be spread over the vibration of several dissimilar bonds fairly evenly and it becomes meaningless to talk of a particular "stretching frequency". For example, approximate normal coordinate calculations on the $PI_3.BI_3$ complex show that the potential energy distribution among the internal coordinates for the band at 303 cm^{-1} (a_1 species) is 12% PB stretch, 13% BI stretch, 20% PI stretch, 16% IPB deformation and 24% IPI deformation (Chantry *et al.* 1966). From the above remarks it is obvious that care must be taken in drawing conclusions about the nature of a given bond in different molecular environments from, say, its "stretching frequency", as the extent to which this vibration will mix with other bond vibrations in the normal mode may vary. While this reservation is made, group frequencies are of limited value and implicit use is made of them for assignments.

C. Phase Effects on Observed Frequencies

This section is necessary in view of the many trends and correlations commonly founded upon frequencies and force constants. Any effects which cause frequency shifts will tend to produce larger percentage differences at low frequencies. Solid-state effects are especially relevant in inorganic systems where vibrational data are often obtained of necessity in the solid state.

1. *Frequency Shifts Associated with Changes of Phase*

When a molecule passes from the gaseous state (where it is essentially an isolated entity) into a condensed phase, there is a change in the potential function, and this will affect the absorption frequencies. This effect is well known in the infrared spectra of organic molecules where, in general, the vibrational frequencies of a given species decrease in the order gas > liquid or solution > solid. However, as mentioned in Chapter 3 (Section XI) at frequencies below 350 cm^{-1} the gas-phase values are lower than the liquid-phase values for a number of organic molecules (Fateley *et al.* 1964). It is not known how general this trend is or if it occurs in any inorganic system. Certainly, in mercuric chloride the stretching frequencies decrease on passing from gas to solution to solid (Allen and Warhurst 1958).

TABLE I. Stretching Frequencies (cm^{-1}) of $HgCl_2$ in Solid, Liquid and Gaseous Phases.

State	ν_1 (Raman)	ν_3 (IR)
Vapour	355	413
C_6H_6 solution	350	392
Solid	316	374

Taimsalu and Wood (1964) in a study of alkyl tin halides found that the Sn–Cl stretching frequencies decreased in the order: nujol > solution > benzene solution > liquid > solid, this being the expected trend from normal experience.

In any event there will be variation between frequencies in the various phases and, hence, comparisons between different species should be confined to the same phase.

2. *Modification of Spectra Due to Solid-State Effects*

The subject of crystal lattice vibrations is dealt with in greater detail in Chapter 8 but a few pertinent points will be briefly mentioned here.

In a crystal lattice it can no longer be assumed that the ion or molecule is an isolated species and the vibrating entity must be considered to be the unit cell. If the symmetry of the species in the unit cell (the site symmetry) is lower than the symmetry of the isolated species (the point group symmetry) there will be a relaxation of the selection rules with a consequent increase in the complexity of the spectrum. For example, any degenerate modes may have their degeneracy lifted with a consequent splitting of the absorption bands. Also, if the species becomes distorted in the crystal lattice, its original point symmetry may be changed and this must be considered together with the site symmetry.

Apart from the above complicating effects another possibility for modification of the low-frequency spectra is the mixing of internal molecular motions and motions of the molecules as a whole (that is, the lattice vibrations) provided they are of the appropriate symmetry. The work of Shimanouchi and co-workers (Shimanouchi and Hiraishi 1966; Shimanouchi and Nakagawa 1962, 1964, 1966a, b; Shimanouchi *et al.* 1964, 1968) is relevant: the complete vibrational spectra (including both internal and lattice vibrations) for a variety of metal-amine and metal halide complexes have been analysed using a modified Urey-Bradley force field. It is found that strong coupling can occur between internal vibrations and lattice vibrations (see also Section IIA1), and such coupling should be more widely considered when normal coordinate calculations are carried out on solid-state systems.

3. *Dielectric Effects*

In condensed phases the true frequency of a vibrational transition is not always the same as the measured frequency of the absorption maximum. This effect is not to be confused with the change in potential function of the species on passing to a condensed phase (*vide supra*) but arises from the variations in the refractive index of a system in the region of an absorption band. This can be envisaged in the following way: a fluctuating electric moment in a molecule makes a contribution to the refractive index. Each normal mode makes a contribution to the refractive index, n, of

$$\Delta_i (n-1) = \frac{N}{6\pi^2 C}\left(\frac{\partial\mu}{\partial Q_i}\right)^2 \Big/ \nu_i^2 - \nu^2$$

where N is the number of molecules per ml, $(\partial\mu/\partial Q_i)$ is the dipole change associated with the normal coordinate Q_i, ν_i is the corresponding frequency of vibration and ν is the frequency of the incident light.

The important features are that if the frequency, ν, of the incident radiation is greater than the transition frequency, ν_i, there is a negative contribution to the refractive index, but if $\nu < \nu_i$ the contribution is positive. The refractive index change will alter the amount of radiation transmitted by the sample apart from any true absorption. Thus, on the high-frequency side the apparent absorption increases, whereas on the low-frequency side it decreases, resulting in a distortion of the band with a shift in the position of the maximum.

In order to analyse a system completely, it is also necessary to measure the reflection spectrum as a function of frequency. In reflection spectros-

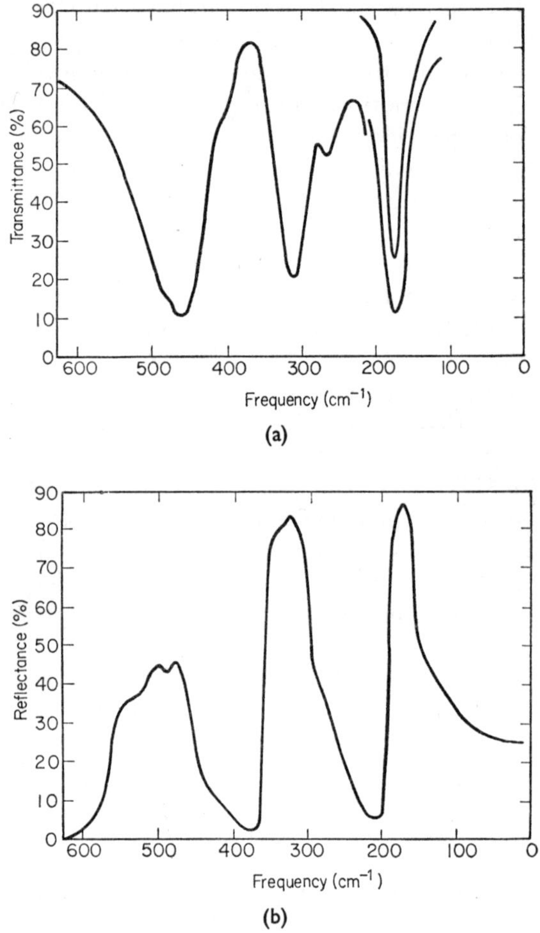

FIG. 1. (a) Transmission, and (b) reflectance spectra of $KMgF_3$.

copy, the reflected energy is dependent on the absorption at this frequency and also on the refractive index of the reflecting medium. From the measured reflectance spectrum, it is possible to determine the true band characteristics (including the true transition frequency) but this involves a complicated analysis of the data using the Kramers-

Kronig relationship and the interested reader is referred to authoritative discussions of the topic (Fahrenfort 1963; Wendlandt and Hecht 1966).

The relevance of this topic is well illustrated in the work of Hunt *et al.* (1964) who investigated the transmittance and reflectance spectra of $KMgF_3$ in the low-frequency region, the spectra being shown in Fig. 1. Analysis of the reflectance spectra by the Kramers-Kronig relation yielded the values of 450, 295 and 140 cm^{-1} for the true transition frequencies, which are seen to be appreciably different from the values of the experimental absorption maxima in the transmittance spectrum (Fig. 1a).

This seems to be an aspect of solid-state measurements which has been widely ignored. However, now that an increasing number of force constant calculations based on frequencies measured in the solid state are being carried out, it should be considered if conclusions about bonding, and so on, are to be based on the derived force constants.

D. Force Constants

Correlation between vibrational frequencies and various properties of molecules and ions are frequently made in inorganic chemistry, these then being employed to make deductions about the nature of related systems.

A more fundamental quantity than the vibrational frequency is the force constant for the stretching of bonds or the deformation of angles between bonds. The usual practice is to carry out approximate calculations on the simpler systems and make correlations between the derived force constants and various structural and electronic properties. A number of such correlations are listed in this chapter and at this stage it is useful to consider the term "bond strength" and define what a force constant measures.

In a diatomic molecule which obeys Hooke's law, the restoring force, F, acting on the system after the bond has been stretched by a small amount, Δx is given by

$$F = -f\Delta x$$

where f is the force constant. The force constant measures the resistance of a bond to an infinitesimally small displacement; thus it provides a measure of the strength of a chemical bond between two atoms in relation to their equilibrium configurations in the molecule. In contrast, the bond dissociation energy, D, does not relate to systems near their potential energy minima. In general, however, the greater the force constant, the greater the bond dissociation energy, and, for the hydrogen halides, a plot of f against D is linear. Many empirical and semi-empirical potential functions exist which relate the two quantities, for example that due to Morse:

$$V = D[1 - \exp^{-a\Delta x}]^2$$

where V is the potential energy, Δx the displacement from the equilibrium bond length and $a = (f/2D)^{1/2}$.

On passing from diatomic to polyatomic molecules, the position becomes rapidly more complicated; not only the stretching of bonds but also the deformation of angles between bonds must now be considered. Also the effects of, for example, the stretching of one bond on the force constant of another become significant and this leads to the introduction of further force constants (interaction constants).

Analogous conclusions about bonding in polyatomic molecules can be made from the derived force constants which, however, are more difficult to evaluate. The problems involved are several and the whole subject is authoritatively reviewed by Mills (1963): only a few of the salient points will be mentioned here. The problems derive largely from the relative shortage of experimental vibrational data. In general, there are more force constants than observed frequencies and in an N-atomic molecule, if the number of normal modes is n, then $n = (3N - 6)$ and in the absence of any symmetry there are $\frac{1}{2}n(n + 1)$ independent force constants in the most general harmonic force field (that is, one in which all the possible interactions are considered). Some extra data from isotopically substituted molecules can be obtained and such quantities as centrifugal distortion and Coriolis coupling constants (both related to force constants) can be extracted from gas-phase vibration-rotation spectra, although this latter type of information is only very rarely available for inorganic molecules. The position is alleviated by the presence of any symmetry in the molecule and high symmetry is often a feature of the simpler inorganic systems such as octahedral MX_6 and tetrahedral MX_4 species. The presence of symmetry reduces the number of independent force constants but even so there are still generally more force constants than vibrational frequencies.

The problem is that of reducing the number of force constants by various assumptions to a solvable number. Various models for such simplified force fields have been proposed but these are nearly all based on the approximation of the valence force field (VFF). In its simplest form this assumes that the only forces are those directed along bonds and angles between bonds, any interactions between, for example, the stretching of different bonds on the bending and stretching of bonds is ignored. This rather oversimplified picture is usually modified in some way to include a number of empirically-chosen interaction terms, such models being called modified valence force fields (MVFF). Thus, considerations which lead to relationships between interaction constants are useful in reducing the number of independent force constants and modified force fields based on this type of reasoning have been used with some success (see also Jones 1962a, b; Mills *vide supra*).

The Urey–Bradley force field (UBFF) and its modifications (MUBFF) have been frequently used to analyse small inorganic systems. Basically, the UBFF involves a simple valence force potential function combined with terms due to interactions between non-bonded atoms in the molecule. The work of Shimanouchi *et al.* is of special interest and includes modifications to the simple UBFF (Shimanouchi 1963), discussion and comparison with general force field constants (Shimanouchi *et al.* 1966), and applications of the field to many inorganic systems (Shimanouchi and Hiraishi 1966; Shimanouchi and Nakagawa 1962, 1964, 1966a, b; Shimanouchi *et al.* 1966, 1968).

The main problem associated with the calculation is choice of model for the force field; a further hazard is the non-uniqueness of solutions to the problem. If two sets of physically plausible force constants both reproduce the observed vibrational frequencies, then it is virtually impossible to decide between them without further experimental data. On the practical level, the shortage of experimental data may be aggravated further by the failure to observe all the fundamental frequencies for reasons such as inactivity in both the infrared and Raman or through inherent weakness of bands.

II. Metal-Ligand Vibrations

A. Metal-Halogen Vibrations

Since fluorine is the lightest halogen and since metal-fluorine bonds are the strongest among the metal halogens, the stretching frequencies are relatively high, and, in general, lie above 400 cm^{-1}. Most metal-chlorine, bromine and iodine vibrations lie below 500 cm^{-1}. Tables of metal-halogen frequencies have been given by Nakamoto (1963) and more recently by Adams (1967). Reviews by Clark (1965a, 1967) are apposite. Some of the trends and correlations which have been drawn will now be mentioned in more detail, but it must always be borne in mind that conclusions drawn from vibrational frequencies alone should be treated with reserve and more definite conclusions must await further normal coordinate calculations. Even where force constants are available, comparisons must be made with great care, bearing in mind the limitations of the various assumptions made in the force field model, and also the other difficulties mentioned above.

1. *Variation of ν(MX) with the Counter Ion in an Ionic Species*

For a particular ion associated with different counter ions, there is usually a shift (often relatively large) in the fundamental frequencies. Many examples can be cited from the literature and those in Tables II and III serve to illustrate the point

TABLE II. Variation of the Infrared Active Frequencies (cm^{-1}) with Cation for Square Planar $[PtCl_4]^{2-}$ Complexes (Sabatini *et al.* 1964).

Cation	$\nu_2(a_{2u})$	$\nu_6(e_u)$	$\nu_7(e_u)$
K	175	325	193
Rb	167	320	190
Cs	160	316	185

TABLE III. Variation of the Infrared and Raman Active Fundamental Frequencies (cm^{-1}) with Cation for Octahedral $[PtCl_6]^{2-}$ Complexes (Adams and Morris 1967a).

			Frequency (cm^{-1})[a]		
Cation	$\nu_1(a_{1g})$	$\nu_2(e_g)$	$\nu_3(t_{1u})$	$\nu_4(t_{1u})$	$\nu_5(t_{2g})$
K^+	350	320	341	186	171
Rb^+	342	316	337	185	173
Cs^+	337	312	332	186	172
Tl^+	353	319	339	182	167
Ag^+	340	328	337	215	186
NH_4^+	345	316	338	193	177
NMe_4^+	339	315	326	182	166
NEt_4^+	340	315	325	181	160
NBu_4^{n+}	335	315	325	181	177
$AsPh_4^+$	337	313	327	187	167

[a] a_{1g}, e_g, and t_{2g} are Raman active; t_{1u} is infrared active.

Many other examples are cited by Adams (1967) and variations over a wide range of cations have been quoted for $[TiCl_6]^{2-}$ (Clark *et al.* 1968b), $[SnCl_6]^{2-}$ and $[PbCl_6]^{2-}$ (Adams and Morris 1967b). In the $[NbCl_6]^-$ ion, however, both the infrared active ν_3 and ν_4 modes show virtually no variation with the cation (Horner *et al.* 1968). It can be seen that some of the anion modes are more sensitive to change of cation than others and, in general, the highest frequencies are found when the anion is associated with the smallest cation. This is undoubtedly a manifestation of the coupling between the internal modes of the complex ion and lattice modes of the crystal. Detailed force constant calculations (using a modified UBFF) on the crystal as a whole have been made by Shimanouchi and Hiraishi (1966) for K_2PtCl_4 and a force constant for the K···Cl interaction in the lattice computed to be 0.115 md/Å. This interaction will vary with the size of the cation and the effective charge on the halogen; in the series K_2PtCl_6, Rb_2PtCl_6 and Cs_2PtCl_6 the following general force field constants were computed.

f (md/Å)	K_2PtCl_6	Rb_2PtCl_6	Cs_2PtCl_6
k_{M-X}	1.86	1.81	1.73
$k_{K\cdots X}$	0.045	0.055	0.068

More complete infrared and Raman studies coupled with further normal coordinate calculations of the above type are plainly required.

Decius (1965) has studied the infrared spectra of KBr doped with $CaSO_4$, $SrSO_4$ and $BaSO_4$. These are consistent with C_{2v} symmetry (8 fundamentals) for the SO_4^{2-} ions indicating the formation of ion pairs in which M^{2+} and SO_4^{2-} ions occupy nearest neighbour sites in the rock salt lattice, thus lifting the degeneracy of the triply degenerate modes of the SO_4^{2-} ion. A model to account for these was developed in terms of the intense electric field due to the excess charge on the alkaline earth ion and relates the observed splitting to the *anharmonic* constants in the potential function of SO_4^{2-}.

2. *Variation of* ν(MX) *with Oxidation State of Metal*

If the coordination number of the metal remains constant, the metal-halogen stretching frequencies generally increase as the oxidation number of the metal increases. This might be expected on the grounds

TABLE IV. Infrared and Raman Fundamental Frequencies (cm^{-1}) for Some Tetrahedral $[Fe^{II}X_4]^{2-}$ and $[Fe^{III}X_4]^-$ Species in the Solid State (Avery *et al.* 1968 and references therein).

Complex	$\nu_1(a_1)$	$\nu_2(e)$	$\nu_3(t_2)$	$\nu_4(t_2)$	f_{M-X} (md/Å) SVFF	MUBFF
$[Et_4N]^+[FeCl_4]^-$	330	114	378	136	2.27	1.51
$[Et_4N]_2^+[FeCl_4]^{2-}$	266	82(?)	286	119	1.48	0.84
$[Et_4N]^+[FeBr_4]^-$	200	—	290	95	1.88	
$[Et_4N]_2^+[FeBr_4]^{2-}$	162	—	219	84	1.24	

that, the greater the positive charge on the M ion, the nearer the negative X ion can approach and the more it will be polarized, thus leading to greater covalent character of the MX bond. This trend is exemplified in Tables IV–VI.

TABLE V. Infrared Active t_{1u} (Antisymmetric MX Stretching ν_3 and ν_4) Frequencies (cm^{-1}) for Some Pairs of Octahedral Haloanions.

Anion	Cation	Oxidation state of metal	$\nu_3(t_{1u})$	$\nu_4(t_{1u})$	Reference
$[NbCl_6]^-$	Cs^+	5	333	165	Horner *et al.* 1968
$[NbCl_6]^{2-}$	Cs^+	4	314	165	Horner *et al.* 1968
$[TaCl_6]^-$	Cs^+	5	328, 317, 307	155	Horner *et al.* 1968
$[TaCl_6]^{2-}$	Cs^+	4	297	160	Horner *et al.* 1968
$[WCl_6]^-$	Cs^+	5	317		Bagnall *et al.* 1964
$[WCl_6]^{2-}$	Cs^+	4	308		Adams *et al.* 1963a
$[IrCl_6]^{2-}$	K^+	4	333	184	Adams and Gebbie 1963
$[IrCl_6]^{3-}$	K^+	3	296	200	Adams and Gebbie 1963
$[UCl_6]^-$	Me_4N^+	5	310		Bagnall *et al.* 1964
$[UCl_6]^{2-}$	Et_4N^+	4	260		Adams *et al.* 1963b

A further example of such a trend is seen in the work of Lewis *et al.* (1965b) who have assigned the infrared-active antisymmetric metal-chlorine stretching frequencies in a series of *o*-phenylene bisdimethyl-arsine (D) complexes.

TABLE VI. Infrared Active Antisymmetric Metal-Chlorine Stretching Frequencies (cm^{-1}) in Some Diarsine Complexes.

	Fe	Ru	Tc	Os	Re
$[M^{III}D_2Cl_2]^+ClO_4^-$	373	340	343	322	325
$M^{II}D_2Cl_2$	349	316	304	288	279

It was pointed out that two reservations should be made about the trend in these pairs, (1) since more than one type of ligand is present, any variation in covalent character due to a change in the charge on the metal ion may be compensated by a change in one type of metal-ligand bond rather than the others, and (2) the bending vibrations in the two types of molecules may be affected by changes in the charge on the metal by different amounts; this could lead to a different degree of interaction between the metal-chlorine stretching and the particular bending vibration in the two cases.

The same effect can be seen in various isoelectronic series of tetrahalides in which the negative charge changes by unity in each member. It is found that, as the negative charge on the complex is increased, the

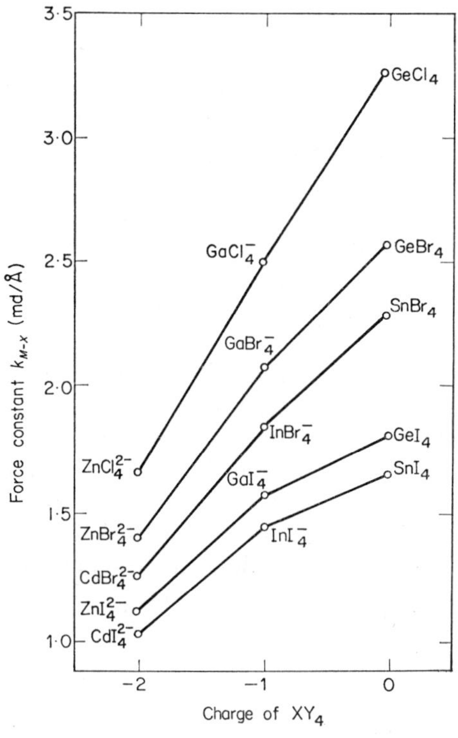

FIG. 2. Variation of force constant k_{M-X} (md/Å) with charge for some isoelectronic tetrahedral species.

metal-halogen stretching frequencies are decreased. This trend is well exemplified in a paper by Woodward (1958) in which the Raman-active totally-symmetric stretching mode, ν_1, is compared among some isoelectronic tetrahedral species. Stretching force constants (using a simple valence force field) were calculated from the ν_1 mode. Frequencies are given in Table VII, and the force constants plotted in Fig. 2.

The same trend is also shown by the infrared active antisymmetric stretching frequencies, ν_3 (cf. Adams 1967 and references therein).

TABLE VII. Raman Active Totally Symmetric Stretching Frequencies, $\nu_1(a_1)$ (cm^{-1}) and Stretching Force Constants, f (md/Å), for Isoelectronic Tetrahedral Species (Woodward 1958).

	X = Cl		X = Br		X = I	
Species	ν_1	$f_{(M-X)}$	ν_1	$f_{(M-X)}$	ν_1	$f_{(M-X)}$
GeX_4	396	3.27	234	2.58	155	1.80
$[GaX_4]^-$	346	2.50	210	2.08	145	1.57
$[ZnX_4]^{2-}$	280	1.64	172	1.39	122	1.11
SnX_4			220	2.28	149	1.66
$[InX_4]^{2-}$			197	1.83	139	1.44
$[CdX_4]^{2-}$			163	1.25	117	1.02

It should be noted that the trend in ν_3 cannot be ascribed to a mass effect, which would operate in the reverse direction, and this is borne out by the force constant values.

3. *Variation of* ν(MX) *with Coordination Number*

In general, for a given oxidation state of a metal, the metal halogen stretching frequencies decrease as the coordination number of the metal increases. This reflects the general tendency for the longer bond lengths in metal complexes with higher coordination numbers. Since the stereochemistry is now being changed within a given series of molecules, care must be taken to specify which stretching mode is being considered; the totally symmetric stretching mode is the obvious choice for this comparison. The following few examples serve to illustrate the trend and other examples are reviewed by Clark (1965a and 1967) and Adams (1967).

(1) The ν_1 modes in the $[BiCl_n]^{3-n}$ systems in aqueous solution taken from a Raman study by Oertel and Plane (1967) are listed below (these frequencies are essentially the same as those observed in the solid complexes).

	$BiCl_3$	$[BiCl_4]^-$	$[BiCl_5]^{2-}$	$[BiCl_6]^{3-}$
Coord. No.	3	4	5	6
ν_1 (cm^{-1})	307	293	280	263

(2) The ν_1 modes in the series $[HgCl_n]^{2-n}$ as observed in the melts containing these species have been assigned by Janz and James (1963a,

b). The Hg–Cl stretching force constants (computed using a Urey-Bradley force field) also followed the same trend.

	$HgCl_2$	$[HgCl_3]^-$	$[HgCl_4]^{2-}$
Coord. No.	2	3	4
ν_1 (cm^{-1})	313	282	267
f_{HgCl} (md/Å)	2.11	1.27	0.82

(3) Woodward and Taylor (1960) have measured the totally symmetric stretching frequencies for the $[InCl_n]^{3-n}$ species as follows:

	$[InCl_4]^-$	$[InCl_5]^{2-}$	$[InCl_6]^{3-}$
ν_1 (cm^{-1})	321	295	275

(4) The $[GeCl_n]^{4-n}$ and $[SnCl_n]^{4-n}$ systems show the same trend.

	$SnCl_4$	$[SnCl_5]^-$	$[SnCl_6]^{2-}$
ν_1 (cm^{-1})	368 [a]	335 [b]	309 [c]
	$GeCl_4$	$[GeCl_5]^-$	$[GeCl_6]^{2-}$
ν_1 (cm^{-1})	397 [a]	348 [b]	318 [c]

[a] Liquid (from Nakamoto, 1963). [b] Solid, as the $Bu^n_4N^+$ salt (Beattie *et al.* 1967b). [c] Solid, as the Et_4N^+ salt (Beattie *et al.* 1967b).

(5) Data on MCl_4, $MCl_4.2L$ and $MCl_4.2D$ systems (where M = Ti and V, L = monodentate ligand and D = *O*-phenylene-bisdimethylarsine) have been summarized by Clark (1965b) and are listed in Table VIII.

TABLE VIII. Variation of ν(TiCl) and ν(VCl) (cm^{-1}) with Coordination Number

Molecule	Coord. No.	ν(M–Cl)
$TiCl_4$ [a]	4	490
$TiCl_4.2L$	6	380
$TiCl_4.2D$	8	317
VCl_4 [a]	4	482
$VCl_4.2L$	6	370
$VCl_4.2D$	8	317

[a] $TiCl_4$ and VCl_4 recorded as liquids; other complexes recorded as solids (nujol mulls).

Very recently, in an attempt to carry out a systematic study of the effect of oxidation state and coordination number on stretching force constants, Wharf and Shriver (1969) have measured far infrared and Raman spectra of a series of complexes containing SnX_6^{2-} (X = Cl, Br and I) and SnX_3^- (X = F, Cl, Br and I) ions. Measurements were all carried out in the solid state, the ions being stabilized by a variety of large cations. Using a modified VFF with a uniform set of approximations, normal coordinate analyses were performed for these complexes and also for SnX_4 (X = Cl, Br and I). The main generalization emerging from this work was that, for a series of complexes containing the same halide, the Sn–X stretching force constant is approximately proportional to the ratio of the oxidation state of the central metal atom to the coordination number.

4. *Variation of* ν(MX) *with Change of* M *and* X

If various series of molecules (for instance a series down, or across, a group in the periodic table) are considered, some trends can be observed in the metal-halogen stretching frequencies, although these will depend on a variety of factors which are not all as yet recognized.

If the metal is varied in any series, the most obvious variation is in the mass of M; on this basis alone, as the mass of M increases, the stretching frequencies will tend to decrease. In such comparisons this effect will always be present and, if possible, it is preferable to eliminate it in order to observe any other trends which may be present. The obvious way to do this is to compare the computed force constants for, say, M–X stretching.

(a) *Variation of* ν(MX) *with change of* X

For a given metal, M, MX frequencies decrease in the order MCl > MBr > MI [see tables in Nakamoto (1963) and Adams (1967)]. This is expected on the basis of a mass effect alone, but apart from this, in general, the bond strengths decrease in this order as well and hence the force constant trend reinforces the mass effect.

Recently an exception to this was claimed by Edwards *et al.* (1968) who used a SVFF to calculate the MnX stretching force constants in the $[MnX_4]^{2-}$ ions (X = Cl, Br, and I) from the Raman-active totally symmetric stretching mode.

The values for the chloride, bromide and iodide are, respectively, 1.31, 1.79 and 1.01 md/Å; that is, a maximum value with the bromide, this being anomalous compared with other known $[MX_4]^{2-}$ ions. It was suggested that the anomaly may be mainly attributable to stretch-stretch interaction terms which would be ignored in a simple VFF calculation. Work by Avery *et al.* (1968) on the Raman spectra of $[MnX_4]^{2-}$ systems in the solid state and in nitromethane solution differs from that of Edwards *et al.* in the assignment of the ν_1 mode for the $[MnBr_4]^{2-}$ species. The assignments and simple VFF Mn–X stretching force constants in both sets of data are summarized below. The 195 cm^{-1} band of Edwards *et al.* is strong and polarized and the 157 cm^{-1} of Avery *et al.* is a weak shoulder with no reported polarization data.

	Edwards *et al.* (1968)		Avery *et al.* (1968)	
	$^{a}\nu_1$ (cm^{-1})	f (md/Å)	$^{b}\nu_1$ (cm^{-1})	f (md/Å)
$[MnCl_4]^{2-}$	249	1.31	258	1.39
$[MnBr_4]^{2-}$	195	1.79	157	1.16
$[MnI_4]^{2-}$	116	1.01	111	0.92

[a] Solution values ($CHCl_3$). [b] Solid state values.

(b) *Variation of* ν(MX) *with Change of* M: *Trends down a group in the periodic table*

Numerous examples can be quoted although explanations of these are not always possible. Large mass changes in passing from one metal to the next complicate the picture and the position will be more satisfactory when further normal coordinate analyses have been performed. A few such trends will be mentioned here.

The series of octahedral hexachloroanions $[TiCl_6]^{2-}$, $[ZrCl_6]^{2-}$ and $[HfCl_6]^{2-}$ have recently been examined by Adams and Newton (1968) and vibrational frequencies and computed force constants are listed in Table IX.

TABLE IX. Vibrational Frequencies (cm^{-1}) and Stretching Force Constants for $[MCl_6]^{2-}$ ions (M = Ti, Zr and Hf).

Anion	Cation	ν_1(R)	ν_2(R)	ν_3(IR)	ν_4(IR)	ν_5(R)	f_{M-Cl}	r_{M-Cl}
$[TiCl_6]^{2-}$	NH_4^+	331	284	330	193	194	1.13	2.31
$[ZrCl_6]^{2-}$	Et_4N^+	327	237	290	150	153	0.88	2.44
$[HfCl_6]^{2-}$	Et_4N^+	333	237	288	145	157	0.86	2.43

These data were all obtained in the solid state and are directly comparable although the $[TiCl_6]^{2-}$ was studied with the NH_4^+ cation, which would result in some shifts relative to the Et_4N^+ cation. This is, however, unlikely to change the order of the frequencies in this series.

The stretching force constants (computed using a modified UBFF) decrease down the group, this being paralleled by increasing M–Cl bond lengths as estimated from Pauling covalent radii. This type of trend is to be expected in the absence of other complicating factors.

The two infrared active M–Cl stretching frequencies of the eight coordinate $MCl_4 2D$ complexes (D = *o*-phenylenebisdimethylamine, M = Ti, Zr and Hf) show a similar variation (Clark 1965b):

$TiCl_4 2D$	317, 325 cm^{-1}
$ZrCl_4 2D$	295, 303
$HfCl_4 2D$	273, 281

It is probable that the MCl stretching-force constants will decrease in the same order in these complexes.

A similar trend is seen in the $VCl_4 2D$ and $NbCl_4 2D$ complexes where the frequencies lie at 317 and 325 for the V complex and 299 and 307 for the Nb complex (Clark 1965b).

Data on the Nb^V and Ta^V octachloroanions appear to be limited to the infrared active ν_3 and ν_4 modes (Adams *et al.* 1963b; Bagnall and Brown 1964; Horner *et al.* 1968). Some of the results of Horner *et al.* for $[NbCl_6]^-$ and $[TaCl_6]^-$ in the solid state with various counterions are listed below.

	K^+		Tl^+		$(C_6H_5)_4As^+$	
Anion	ν_3	ν_4	ν_3	ν_4	ν_3	ν_4 (cm^{-1})
$[NbCl_6]^-$	333	176, 163	335	165	331	163
$[TaCl_6]^-$	315	154	333	164	316	155

For the counterions K^+ and $(C_6H_5)_4As^+$, there is a decrease on passing from Nb to Ta but caution should be exercised in drawing conclusions, for example it is seen that there is virtually no difference in the frequencies for the Tl^+ cation and this is also the case for the Et_4N^+ cation (Adams *et al.* 1963b). The ν_3 and ν_4 frequencies (cm^{-1}) in some tetravalent Nb and Ta octahaloanions are listed below (Horner *et al.* 1968). These show a reduction on passing from Nb to Ta but it will be difficult to make any satisfactory deductions about relative bond strengths until a complete Raman and infrared study is carried out and force constants computed.

Complex	ν_3 (cm^{-1})	ν_4 (cm^{-1})
$Cs_2[NbCl_6]$	314	165
$Cs_2[TaCl_6]$	297	160
$Cs_2[NbBr_6]$	236	112
$Cs_2[TaBr_6]$	217	109

For the $[WCl_6]^{2-}$ and $[MoCl_6]^{2-}$ ions, the infrared-active antisymmetric stretch (ν_3) is greater for Mo than W for several cations, typical values being 325 cm^{-1} and 308 cm^{-1} for $Cs_2[MoCl_6]^{2-}$ and $Cs_2[WCl_6]^{2-}$, respectively (Adams *et al.* 1963a). In complexes of the type $Et_4N^+[M(CO)_5Cl]$, where M = Cr, Mo and W, the infrared-active MCl stretching frequencies are 257 cm^{-1} in $[Cr(CO)_5Cl]^-$, 248 cm^{-1} in $[Mo(CO)_5Cl]^-$ and 258 cm^{-1} in $[W(CO)_5Cl]^-$ (Bennett and Clark 1964). From these values it might be inferred that the WCl bond is the strongest of the three in these systems, as the reverse trend should obtain on the basis of the mass effect.

In $Mn(CO)_5Cl$ and $Re(CO)_5Cl$ the MnCl and ReCl stretching modes lie at 295 and 294 cm^{-1}, respectively, thus suggesting that the latter may have the larger force constant. The corresponding values for the bromide complexes are 218 cm^{-1} (MnBr) and 202 cm^{-1} (ReBr) (Bennett and Clark *vide supra*). In the $Tc^{II}Cl_2D_2$ and $Re^{II}Cl_2D_2$ complexes (D = *o*-phenylenebisdimethylarsine) the infrared-active M–Cl stretches lie at 304 and 279 cm^{-1}, respectively (Lewis *et al.* 1965b). In the M^{III} complexes $[M^{III}D_2Cl_2]^+ClO_4^-$ the infrared-active TcCl and ReCl stretches lie at 335 cm^{-1} and 313 cm^{-1}, respectively.

The infrared-active ν_3 mode for $[Ru^{IV}Cl_6]^{2-}$ and $[Os^{IV}Cl_6]^{2-}$ were observed at 346 and 328 cm^{-1}, respectively, for the potassium salts (Adams and Gebbie 1963), and in the complexes $M^{II}D_2Cl_2$ and $[M^{III}D_2Cl_2]^+ClO_4^-$ (for M = Fe, Ru, Os) there is a decrease in the MCl stretching frequency (cm^{-1}) on passing down the group (Lewis *et al.* 1965b), as shown below.

M	$M^{II}D_2Cl_2$	$[M^{III}D_2Cl_2]^+ClO_4^-$	
Fe	349	373	(cm^{-1})
Ru	316	340	
Os	288	322	

However, in the *cis* $Fe(CO)_4Cl_2$, the two infrared-active FeCl stretching modes lie at 294 and 318 cm^{-1} and the corresponding modes in

$Os(CO)_4Cl_2$ at 298 cm^{-1} and 328 cm^{-1}, thus suggesting that the bond strength might be greater in the third row element in this series (Bennett and Clark, 1964).

In $[RhCl_6]^{3-}$, ν_3 lies at 322 cm^{-1} (Shimanouchi *et al.* 1964) and at 296 cm^{-1} in $[IrCl_6]^{3-}$ (Adams and Gebbie 1963). For the $[M^{III}D_2Cl_2]^+ClO_4^-$ complexes (Lewis *et al.* 1965b), the MCl stretching mode decreases in the order CoCl (388 cm^{-1}), RhCl (358 cm^{-1}), IrCl (335 cm^{-1}). In the square planar Rh^I and Ir^I complexes, $Rh(CO)Cl(PPh_3)_2$ and $Ir(CO)Cl(PPh_3)_2$, RhCl and IrCl stretching modes were observed at 309 cm^{-1} and a pair at 317 and 311 cm^{-1}, respectively, suggesting that some bond strengthening might occur on passing from Rh to Ir in these systems (Bennett *et al.* 1967).

Many data exist for the Ni, Pd and Pt series. Coates and Parkin (1963) have studied a wide range of square planar *trans* MX_2L_2 species (where M = Ni, Pd and Pt and L = other ligands), typical MX stretching frequencies being listed below. This is the same order observed for the

$NiCl_2[P(CH_3)_3]_2$	$PdCl_2[P(CH_3)_3]_2$	$PtCl_2[P(CH_3)_3]_2$
403 cm^{-1}	355 cm^{-1}	326 cm^{-1}

ν_3 mode of $[PdCl_6]^{2-}$ and $[PtCl_6]^{2-}$ ions, typical values (for the Et_4N^+ cation) being 337 cm^{-1} and 330 cm^{-1} (Adams *et al.* 1963b). However, Shimanouchi *et al.* (1964) using a UBFF computed MX stretching force constants of 1.51 and 1.86 md/Å, respectively, for $[PdCl_6]^{2-}$ and $[PtCl_6]^{2-}$, indicating the PtCl bond to be the stronger.

In the series $[MCl_4]^{2-}$ (where M = Zn, Cd and Hg) the ν_3 infrared-active antisymmetric stretching modes occur at 277 cm^{-1} (Sabatini *et al.* 1964), 260 and 228 cm^{-1}, respectively (Adams *et al.* 1963b). For the $[MI_4]^{2-}$ series of Zn, Cd and Hg, Raman data were used to compute force constants using a Urey–Bradley–Shimanouchi force field (Long and Chau 1962 and references therein). The fundamental frequencies (cm^{-1}) of these are listed below.

	$[ZnI_4]^{2-}$	$[CdI_4]^{2-}$	$[HgI_4]^{2-}$
$\nu_1\ (a_1)$	122	117	126
$\nu_2\ (e)$	44	36	35
$\nu_3(t_2)$	170	145	—
$\nu_4\ (t_2)$	64	44	41

With the exception of ν_1 (Raman-active, totally-symmetric stretch) the frequencies all decrease in the order Zn > Cd > Hg. The computed force constants (in md/Å) for MI stretching were found to decrease in the order:

$$HgI_4^{2-}\ (0.83) > CdI_4^{2-}\ (0.64) > ZnI_4\ (0.58)$$

that is, the opposite direction to that which might be inferred from the ν_3 frequencies.

It may be noted that on the simple valence force field model the MX stretching force constant could be obtained directly from a knowledge of the Raman-active (a_1) frequency, ν_1, and for a given X group the force constant is proportional to the square of ν_1. Hence, inspection of the ν_1 modes listed above shows that a different order of stretching force constants would be obtained on a SVFF approximation. In this particular case the SVFF would probably be a gross over-simplification since non-bonded interactions (due to the large iodine atoms) are almost certainly important. However, this illustrates how force constant trends could be misleading.

In the group IV hexachloroanions, a general decrease in the infrared-active frequencies (ν_3 and ν_4) is seen on passing down the group, as shown below.

	ν_3 (cm^{-1})	ν_4 (cm^{-1})	f (md/Å)	References
$Et_4N^+[GeCl_6]^{2-}$	310	213		Beattie *et al.* (1967)
$NH_4^+[SnCl_6]^{2-}$	310	180	1.08	Adams and Morris (1967b)
$NH^+[PbCl_6]^{2-}$	278	154	0.86	Adams and Morris (1967b)

The force constants for $[SnCl_6]^{2-}$ and $[PbCl_6]^{2-}$ were computed using a modified UBFF, and suggest that the SnCl bonds are stronger than the PbCl bonds.

(c) *Variation of* ν(MX) *with Change of* M *Across the Periodic Table*

When trends of this type are considered for transition metals, there are not the large percentage mass differences between each member as in series down a group. The infrared spectra of the $M^{II}X_4^{2-}$ species (M = Mn, Fe, Co, Ni, Cu, and Zn; X = Cl, Br and I) have been studied by various workers (Adams *et al.* 1963b; Clark and Dunn 1963; Sabatini *et al.* 1964), the antisymmetrical stretching frequency ν_3 being assigned in each case. The variation of ν_3 for the series of tetraalkylammonium salts of MX_4^{2-} is summarized in Fig. 3 (Clark 1965a). The ν_3 frequencies

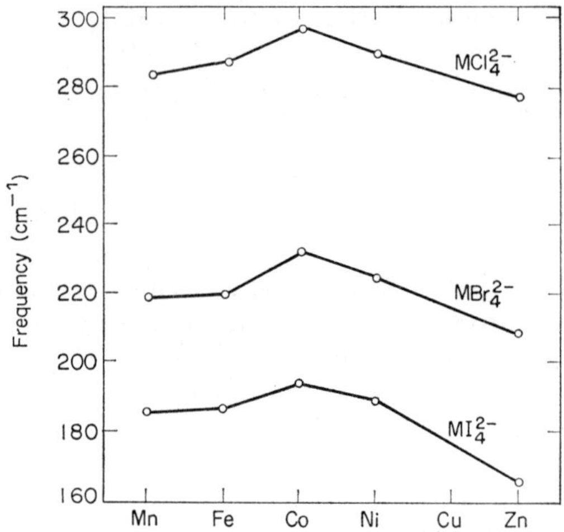

FIG. 3. Variation of ν_3 (cm^{-1}) in $[MX_4]^{2-}$ species with change of M.

follow the same pattern for chlorides, bromides and iodides, namely Mn $\leqslant$ Fe $<$ Co $>$ Ni $>$ Zn which corresponds to the order of ligand field stabilization energies. (The copper complexes were not included in the diagram because of appreciable distortion from tetrahedral symmetry.) Very recently some MX stretching force constants in the series $(Et_4N^+)_2[M^{II}X_4]^{2-}$ (M = Mn, Fe and Zn; X = Cl, Br and I) were computed using a valence force field from solid-state Raman data (Avery *et al.* 1968); these are summarized below.

	f_{M-Cl} (md/Å)		f_{M-Br} (md/Å)
$[MnCl_4]^{2-}$	1.39	$[MnBr_4]^{2-}$	1.16
$[FeCl_4]^{2-}$	1.48	$[FeBr_4]^{2-}$	1.24
$[ZnCl_4]^{2-}$	1.59	$[ZnBr_4]^{2-}$	1.38

These data suggest that the strongest bonds occur at Zn, which is in contrast to the order suggested by the ν_3 frequencies. Further force constant computations on these and other series of complexes are plainly required.

A similar trend can be seen in the infrared-active MCl stretching frequencies in octahedral complexes of the type $[M^{III}Cl_2D_2]^+$ (where D = *o*-phenylenebisdimethylarsine) (Lewis *et al.* 1965b).

The maximum at Co^{III} is followed by a very sharp fall to 260 cm^{-1} in the Ni^{III} complex. The d^6 configuration corresponds to the maximum

TABLE X. I.R.-Active M–Cl Stretching Frequencies (cm^{-1}) in $[M^{III}D_2Cl_2]^+$ Complexes.

M^{III}	d^n	ν(MCl)
Cr	d^3	375
Fe	d^5	373
Co	d^6	388
Ni	d^7	260

ligand field stabilization energy for octahedral systems and on passing from d^6 (that is, t_{2g}^6) to d^7 (that is, $t_{2g}^6 e_g^1$) the sharp drop in the frequency is explained by the antibonding nature of the e_g orbital into which the seventh electron is added.

Very recently Adams and Morris (1968) have observed the same effect in the high-spin series $[M^{III}Cl_6]^{3-}$ using the cation [Co(1,2 diaminopropane)$_3]^{3+}$ where M = Cr, Mn and Fe. Frequencies (cm^{-1}) of the infrared-active (ν_3 and ν_4) modes are listed below.

	d electron config.	ν_3	ν_4
$[CrCl_6]^{3-}$	t_{2g}^3	315	200
$[MnCl_6]^3$	$t_{2g}^3 e_g^1$	342	183
$[FeCl_6]^{3-}$	$t_{2g}^3 e_g^2$	248	184

The drastic change of the antisymmetric stretching frequency, ν_3, on passing from Cr^{III} to Fe^{III} is explained in terms of the occupation of the antibonding e_g orbitals in the Fe^{III} complex (the $[MnCl_6]^{2-}$ system is complicated by effects of Jahn-Teller distortion).

The MCl stretching force constants (obtained using a MUBFF) in the complexes $[OsCl_6]^{2-}$, $[IrCl_6]^{2-}$ and $[PtCl_6]^{2-}$ were found to be 1.5, 1.6 and 1.86 md/Å, respectively (Shimanouchi *et al.* 1964). This follows the same order as the CFSE for low-spin $[MCl_6]^{2-}$ complexes.

5. *Effect of Variation of Other Ligands on* ν(MX)

(a) *Variation of the Relative Numbers of Halogens and Other Ligands for a Given Metal*

The best example of this trend afforded by low-frequency spectroscopy is the tin–halogen stretching frequencies in the series $R_{4-n}SnCl_n$, data for which are summarized by Clark *et al.* (1968a). In general, the tin–chlorine stretching frequencies decrease with increasing alkylation, and in Fig. 4 the degeneracy-weighted average SnX (X = Cl, Br and I) stretching frequency is plotted against the degree of alkylation for the methyl derivatives. The average values of the SnX stretching frequencies in the ethyl and *n*-butyl tin-halides are very similar to those of the analogous methyl tin-halides. It is believed that, as the chlorine atoms are replaced by the electron-releasing alkyl groups, the effective nuclear charge on the tin atom is reduced, with consequent weakening of the remaining tin–chlorine bonds. This argument is supported by the increasing tin–halogen bond lengths with increasing alkylation.

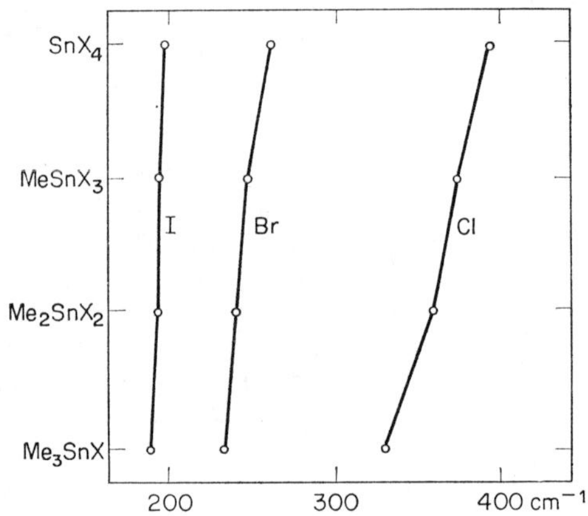

FIG. 4. Variation of the degeneracy-weighted Sn–X stretching frequencies (cm^{-1}) in the series $(CH_3)_{4-n}SnXn$.

(b) *Variation of* L *in* MX_nL *Systems*

In this type of series the MX_n skeleton remains the same and only the ligand(s) L varies. Several investigations of platinum and palladium halide complexes have been carried out and correlations made between the nature of L and MCl stretching frequencies.

Adams *et al.* (1964) studied several series of square planar PtCl and PtBr complexes. For the series *trans* PtX_2L_2, the PtX stretching frequencies were almost insensitive to the nature of L, falling in the range 339.5 ± 3.1 cm^{-1} when X = Cl, and 249.8 ± 9.2 cm^{-1} when X = Br; that is, the ligands L have little effect on the bonds in the position *cis* to the metal-ligand bond. In the series of *trans* $[PtXR(PEt_3)_2]$ (X and R *trans* to one another) and in the *cis* isomers of PtX_2L_2 (L and X *trans* to one another), the PtX stretching frequencies are very dependent on the nature of L. In general, it was found that the most important factor determining the PtX bond stretch was the electro-negativity of L, but also anomalies could be explained by π-bonding in the Pt–L bond. It was pointed out that, although there may be coupling to a different extent for different ligands between Pt–L and Pt–X vibrations, the fact that the same order was observed in both Cl and Br complexes indicates that a *trans* effect was being observed rather than an effect simply due to coupling. Hence, it is reasonable to make deductions about the effect of L on the Pt–X bond strength. Similar trends can be seen in the series of *cis* PdX_2L_2 (Coates and Parkin 1963).

A series of $M_2Cl_4L_2$ complexes (M = Pt and Pd, L = ligand) has been studied by Goodfellow *et al.* (1967). These were assigned on the basis of the structure below. The M–Cl bridge bond (that is, *trans* to M–L)

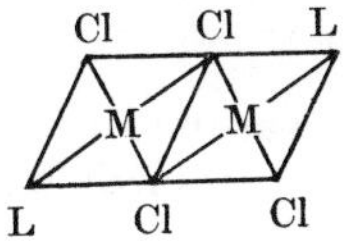

stretching frequencies, in general, followed the same trend as found by Adams *et al.* (1964) for the *cis* $[PtCl_2L]$ series.

The effect of variation of ligand L on the MX stretching frequencies in various Rh^{I} and Ir^{I} halogeno complexes has been investigated by Bennett *et al.* (1967) together with octahedral complexes of Rh^{III} and Ir^{III}. It was found that the MX stretching frequencies were mainly dependent on the ligand in the *trans* position to the MX bond.

Further examples of $SnCl_4L_2$ (Beattie *et al.* 1963; Beattie and Rule 1964) and MX_2L_2 where M = Zn, Cd and Hg (Coates and Ridley 1964) serve to illustrate this point.

6. *Halogen Bridging Stretching Frequencies*

Species which contain metal atoms linked by halogen bridges are a common feature in inorganic chemistry, and range from the simple Al_2X_6 systems to complex "cluster" compounds; the latter are considered separately in Section IV.

As yet, detailed vibrational data on halogen bridged systems are relatively sparse. Vibrational spectra are of relevance to the study of such systems, (a) to obtain force constants and, hence, information about the strength of the bonds, and (b) as a possible diagnostic test for the presence of the bridging bonds in unknown systems.

Without doubt the Al_2X_6 systems have attracted the most detailed interest (see Adams and Churchill 1968 and references therein). In these systems the aluminium atoms are in a tetrahedral environment, and the two aluminium atoms and the two bridging chlorines are coplanar.

It has been established (Onishi and Shimanouchi 1964) that the bridge bond stretching-force constant in Al_2Cl_6 is about half the value of that for the terminal bonds. Beattie *et al.* (1967a) obtained a good fit of calculated and observed frequencies in Ga_2Cl_6 by using a force constant for the terminal GaCl stretching of about twice that for the bridging GaCl stretch.

In such M_2X_6 systems of point group D_{2h} (see Table XXXI) there should be two modes active in the infrared, mainly associated with terminal MX stretching (b_{1u} and b_{3u}) and two mainly associated with bridge MX stretching (b_{2u} and b_{3u}). The values which have been assigned to these modes for various group III dimeric halides are collected in Table XI.

TABLE XI. Infrared-Active Terminal and Bridge Stretching Frequencies in Some Group III Dimeric Halides.

	$\nu(MX)_t$ (cm^{-1})		$\nu(MX)_b$ (cm^{-1})		
	b_{1u}	b_{3u}	b_{2u}	b_{3u}	References
Al_2Cl_6	625	484	420	(301)	Adams and Churchill (1968)
Al_2Br_6	500	373	341	198	Adams and Churchill (1968)
Al_2I_6	420	334	296	143	Adams and Churchill (1968)
Ga_2Cl_6	477	400	310	280	Beattie *et al.* (1967a)
Ga_2Cl_6	420	394	365	466	Balls *et al.* (1966)
Ga_2Br_6	346	313	227	264	Balls *et al.* (1966)
Ga_2I_6	270	250	185	210	Balls *et al.* (1966)
In_2I_6	228	178	158	125	Adams and Churchill (1968)
In_2I_6	214	—	153	173	Greenwood *et al.* (1968)

In the aluminium halides, the halogen bridging modes generally lie at lower frequencies than the terminal modes but there are some disagreements among the assignments observed for the gallium and indium systems. Assignments for some other tetrahedral dimeric MCl_3 halides have been reported. The terminal FeCl stretches in Fe_2Cl_6 were assigned at 507 and 399 cm^{-1} and a bridging mode at 268 cm^{-1} (Wilmshurst 1960). Coates and Ridley (1964) examined the spectra of a number of bipyridyldiphenylboronium (B^+) salts of the type $B^+MX_3^-$ (where M = Zn, Cd and Hg and X = Cl and Br). It was suggested that these are probably $[M_2X_6]^{2-}$ ions. In the $[Zn_2Cl_6]^{2-}$ spectra, four strong bands were observed, the two higher frequencies being assigned to terminal and the lower two to bridging modes; these data are summarized below.

	$[Zn_2Cl_6]^{2-}$	$[Zn_2Br_6]^{2-}$	$[Cd_2Cl_6]^{2-}$
ν (Terminal) (cm^{-1})	334	253	271
	301	209	224
ν (Bridging) (cm^{-1})	238	<200	<200
	225		

Another group of halogen bridged molecules for which some vibrational data are now available is that in which the metal atoms have a square planar coordination, as shown below. The point group for such systems

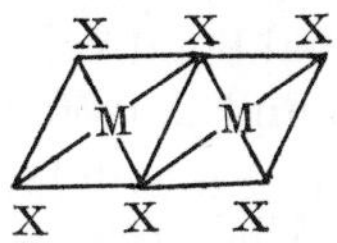

is again D_{2h} and four stretching modes mainly associated with the bridging MX system and four mainly associated with the terminal MX groups are expected (see Table XXXII for distribution of modes and activities).

The best known examples of ions of this type are the $[Pt_2X_6]^{2-}$ and $[Pd_2X_6]^{2-}$ systems where X = Cl, Br and I. Adams *et al.* (1967) have analysed the infrared spectra of these systems and, in rough calculations, found that the MCl bridge bond stretching force constants ranged from 75–100% of those for the terminal ones, that is, a considerably higher ratio than in the case of the Al_2Cl_6 and Ga_2Cl_6 systems. The frequencies are summarized in Table XII.

Very recently, Adams and Churchill (1968) in a more detailed investigation and force constant calculation on Au_2Cl_6 have obtained the values

TABLE XII. Vibrational Frequencies (cm^{-1}) for Infrared-Active Terminal and Bridge Stretching Modes in $[M_2X_6]^{2-}$ Systems.

	$[Pd_2Cl_6]^{2-}$	$[Pd_2Br_6]^{2-}$	$[Pd_2I_6]^{2-}$	$[Pt_2Cl_6]^{2-}$	$[Pt_2Br_6]^{2-}$	$[Pt_2I_6]^{2-}$
Cation	$Bu_4^nN^+$	Et_4N^+	Et_4N^+	Ph_4As^+	Et_4N^+	Et_4N^+
$\nu(MX)_t$						
b_{2u}	340	266	222	346	237	179
b_{3u}	349	262	220	350	243	194
$\nu(MX)_b$						
b_{2u}	302	192	144	315	196	157
b_{3u}	264	177	132	302	211	144

TABLE XIII. MX Stretching Force Constants (md/Å) for $[M_2X_6]^{2-}$ (M = Pd and Pt, X = Cl, Br and I).

M	Pd		Pt		
X	Cl	Br	Cl	Br	I
k_t	1.59	1.37	2.02	1.62	1.29
k_b	1.36	1.30	1.42	1.56	1.29

of 2.38 and 1.74 md/Å for the stretching force constants of the terminal and bridging bonds respectively; that is, the bridge stretching force constant is 73% of the terminal bond value. It will be interesting to see if the values of $(k_t/k_b) \simeq 0.75$ and $\simeq 0.50$ are obtained for other square planar (for example $[Cu_2Cl_6]^{2-}$) and tetrahedral (for example Fe_2Cl_6, In_2Cl_6) systems, respectively, when they have been more fully analysed. It is suggested by Adams and Churchill (*vide supra*) that, in the square planar systems, the sharing of an edge results in little distortion of the metal–metal distance, whereas, if the two tetrahedra share an edge the metal atoms are brought more closely together which leads to electrostatic repulsion and a consequent lengthening of the bridge bonds. Detailed studies on more systems of both types are obviously required, but the evidence to date suggests that the force constants and stretching frequencies for terminal MX bonds will generally be greater than those for the bridged part of the system.

Apart from these more detailed studies, a number of systems have been studied from the viewpoint of stretching frequencies alone. The series of complexes of the type $Rh_2X_2L_4$ in which a square planar

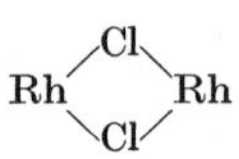

TABLE XIV. Infrared-Active RhX Stretching Frequencies in Some $Rh_2X_2L_4$ Complexes.

	ν_{bridge} (cm^{-1})	
Rh_2Cl_2 (cyclooctadiene)$_2$	277.9	260.2
Rh_2Cl_2 $(C_2H_4)_2$	275.7	262.1
Rh_2Cl_2 $(SnCl_3)_4$	288.4	271.6
Rh_2Br_2 (cyclooctadiene)$_2$	194.8	184.4
Rh_2Br_2 $(SnBr_3)_4$	197.5	175.8

bridge is involved have been investigated by Adams and Chandler (1965). In these systems the spectra are uncomplicated by terminal RhCl frequencies and assignment is thus reasonably certain. The two infrared-active bridge stretching modes are assigned in Table XIV.

A number of correlations for various Pd^{II} and Pt^{II} complexes of the types $[M_2X_6]^{2-}$ and $M_2X_4L_2$ for X = Cl, Br and I and L = PCl_3, $POEt_3$, PR_3, AsR_3, SR_2, SeR_2, TeR_2, pyridine and olefins have been made by Adams and Chandler (1966). For the $M_2X_4L_2$ (and including most $[M_2X_6]^{2-}$ systems) the following ranges were observed in the far infrared (see Table XV).

TABLE XV. Terminal and Bridge Stretching Frequencies (cm^{-1}) in $M_2X_4L_2$ Systems (M = Pd, Pt; X = Cl, Br and I; L = ligand).

M	X	Terminal $\nu(MX)$[a]	Bridging $\nu(MX)$
Pt	Cl	365–340	335–310, 295–220
Pt	Br	260–235	230–210, 190–170
Pt	I	200–170	190–150, 150–135
	X		
Pd	Cl	370–345	310–300, 280–250
Pd	Br	285–265	220–185, 200–165

[a] The two infrared active terminal MX stretching frequencies usually lie close together (within 5 cm^{-1}) and only one range is quoted to include both of these.

The complexes $KCuCl_3$, NH_4CuCl_3 and $LiCuCl_32H_2O$ all contain the planar $[Cu_2Cl_6]^{2-}$ ion. Adams and Lock (1967) have assigned the terminal and bridging frequencies in these species (Table XVI) and again the two

TABLE XVI. Infrared-Active CuCl Terminal and Bridge Stretching Frequencies (cm^{-1}) in Square Planar Halogen-Bridged Systems.

	ν_{C_uX} (terminal)	ν_{C_uX} (bridge)
$Li_2[Cu_2Cl_6]^{2-}$	295	272, 222
$NH_4[Cu_2Cl_6]^{2-}$	311	280, 230
$K_2[Cu_2Cl_6]^{2-}$	301	278, 236
Cu_2Cl_4 (pyridine-1-oxide)$_2$	325 (sh, 315)	—
$K_2[Cu_2Br_6]^{2-}$	237	

terminal modes lie very close together. In the complex Cu_2X_4 (pyridine-1-oxide)$_2$ the four halogens are all in terminal positions, the copper atoms being bridged by the pyridine-1-oxide molecules. In this complex a single band at 315 cm^{-1} with a shoulder at 325 cm^{-1} must be the terminal CuCl stretching frequency. This strongly supports the assignment of the higher frequency band in $[Cu_2Cl_6]^{2-}$ to the CuCl terminal stretch.

B. Complexes Containing Ligands with Heavy Donor Atoms

A large volume of low-frequency vibrational data has now been accumulated for systems involving ligands which contain non first-row donor atoms, especially phosphorus and sulphur and, to a lesser extent, arsenic, selenium and tellurium; however, these data are not as well documented as metal-halogen frequencies. Whereas the ligands are simple monatomic halide ions in the metal-halogen complexes, the heavy donor atoms of groups V and VI are usually part of a complicated ligand molecule which has its own internal vibrations. This complicates the assignment in that frequencies arising from modes of the metal ligand skeleton have to be distinguished from the low-frequency internal ligand modes. There will also be coupling between metal ligand vibrations and internal ligand vibrations of the appropriate symmetry, thus modifying the frequencies and band intensities found for the ligand itself. The likelihood of such coupling is increased by the presence of a heavy atom in the ligand, which will lead to low-frequency internal ligand modes. If chelating ligands are involved, the position is complicated even further by the presence of ring modes which are not present in the parent ligand. Often the ligand itself is a liquid but its complexes are solids and must often be examined as such. Space does not permit a comprehensive survey of the work in this field but the position up to the end of 1965 has been reviewed by Adams (1967) and a few examples chosen from the literature since this time will be given here.

One of the future trends in investigations of this type of system will probably be more study in depth of series of closely related systems; that is, detailed infrared and Raman data coupled with normal coordinate calculations. One of the prerequisites of such work will be more detailed investigations of the ligands themselves in order to facilitate the interpretation of the spectra of the complexes. A greater number of this type of study is now being made.

1. *Sulphur, Selenium and Tellurium*

Most of the work on vibrational spectra of complexes with sulphur- and selenium-containing ligands has involved various diethyl sulphides and selenides, thiocyanates and selenocyanates (see Adams 1967 and references therein) and, more recently, thiourea and selenourea (see, for example, Adams and Cornell 1967; Hendra and Jovic 1968).

In some of this recent work (Adams and Cornell 1967) a systematic study of thirty-six thiourea and ethylene thiourea complexes was made, and metal sulphur vibrational frequencies assigned. The variation of these with stereochemistry and coordination numbers was demonstrated: for example, the Co–S stretch in the six-coordinate $CoCl_2$ (thiourea)$_4$ was assigned at 205 cm^{-1}, well below the range 298–250 cm^{-1} for the four-coordinate [Co(thiourea)$_4$]^2 and CoX_2(thiourea)$_2$ series.

Infrared and Raman spectra of square planar complexes of the type [M^{II}(thiourea)$_4$]X_2 (where M = Pd and Pt and X = Cl) have been studied by Hendra and Jovic (1968). The metal ligand modes have been assigned on the basis of a planar pseudo-pentatomic skeleton. On this basis, the totally symmetric M–S stretching modes (a_{1g}) were assigned to the polarized Raman lines as follows.

	ν_1 (cm^{-1})
[Pd(thiourea)$_4$]$^{2+}$	257
[Pt(thiourea)$_4$]$^{2+}$	273
[Pd(selenourea)$_4$]$^{2+}$	178
[Pt(selenourea)$_4$]$^{2+}$	191

Allkins and Hendra have carried out a systematic study of infrared and Raman spectra of ligands of the type $(CH_3)_2Y$, where Y = S, Se and Te (1966, 1967a), the effect of complex formation on the internal modes of these ligands (1968) and the low-frequency metal-ligand modes of various *cis*- and *trans*- $MX_2(Me_2Y)_2$ and $[MX_2(Me_2Y)]_2$ bridged platinum and palladium complexes (1967b). In general for these complexes the assigned metal-chalcogen stretching frequencies (cm^{-1}) were approximately as follows.

Pt–S (sym)	345
Pt–S (a.s.)	312
Pt–Se (sym)	175
Pt–Se (a.s.)	230
Pt–Te (sym)	165
Pt–Te (a.s.)	230

These frequencies were relatively unchanged on passing from Pt to Pd. The order of frequencies is PtS > PtSe < PtTe and calculation of very approximate force constants (taking the ligand as a point mass) leads to values of $k_{Pt-S} = 2.2$, $k_{Pt-Se} = 1.4$ and $k_{Pt-Te} = 2.1$ md/Å which obviously has no absolute significance but may indicate a trend in the bond strengths.

Some detailed infrared studies coupled with normal coordinate analysis have been made on sulphur-containing chelate systems. Dithiocarbamato

(Nakamoto *et al.* 1963), and dithiooxalato (Fujita and Nakamoto 1964) complexes of platinum have been studied in this way. The dithiocarbamato complexes of several other metals besides Pt were examined, but no normal coordinate analysis was carried out. The modes previously involving M–S stretching were shifted to high frequencies as the metal was changed in the order Pt^{II}, Pd^{II}, Ni^{II}, Co^{III}, Cr^{III} and Cd^{II}, but it was pointed out that no deductions about the M–S bond strength could be made purely on this order of frequencies.

2. *Phosphorus and Arsenic*

Trialkyl and triaryl phosphines, together with phosphorus trihalides, are the commonest P-containing ligands of the numerous Pt and Pd complexes which have been examined (see Adams 1967 and references therein).

More recently, for example, far-infrared spectra of Pt^{IV} complexes of the type *cis* and *trans* PtX_4L_2 (X = Cl, Br and I, L = donor ligand, among which are PEt_3 and $AsEt_3$) have been examined (Adams and Chandler 1967). Assignments are given for the Pt–P and Pt–As stretching modes. Although arsenic-containing ligands have played an important part in stabilizing the less commonly occurring oxidation states of metals, few systematic low-frequency investigations have been made. Several M–As stretching frequencies have been assigned by Coates and Parkin (1963) in some gold, palladium and platinum complexes with trimethylarsine, and these fall in the range 240–280 cm^{-1}. Recently, Green and his collaborators have published detailed investigations of various important ligands together with studies on the complexes of some of these. Deacon and Green (1968a) have reported on the spectra of triphenylphosphine and various complexes of it, especially those with MX_2 (M = Zn, Cd and Hg; X = Cl, Br and I), although this investigation only extended down to 375 cm^{-1}. The arsenic-containing ligands, dimethylphenylarsine, methyldiphenylarsine, triphenylarsine and *o*-phenylbisdimethylarsine have also been investigated down to 200 cm^{-1} by Green *et al.* (1968). Complexes of *o*-phenylenebisdimethylarsine (D) of the type MDX_2 (M = Zn, Cd and Hg; X = Cl and Br) were investigated by Deacon and Green (1968b) down to 200 cm^{-1} together with those of the type $[MD_2]^{2+}(ClO_4)_2$.

Many complexes of ligands involving heavy donor atoms have been prepared and, as more and more vibrational spectra are measured and conclusions drawn from, say, M–X modes, the greater will be the need for more detailed data on the ligands, and the effect on these of complex formation, in order to assist assignment of the metal-ligand skeletal modes. For more reliable conclusions to be drawn about the bonding in such systems there is little doubt that the more detailed infrared and Raman measurements are required together with the force constant calculations.

III. Metal–Metal Stretching Frequencies

In recent years much work has been directed towards the preparation of compounds containing metal–metal bonds and this, coupled with the recognition of such bonds in previously known compounds, has led to considerable interest in their properties and structure, including vibrational spectra. It has been suggested that it might be possible to use characteristic ranges of such frequencies as a criterion of the presence of given metal–metal bonds, although this would seem to be an optimistic view.

Apart from the stretching vibrations and force constants for many heteronuclear and homonuclear diatomic metals in the gaseous state which are tabulated by Herzberg (1950), the first metal–metal stretching frequency was recognized by Woodward (1934) using Raman spectroscopy. The band at 169 cm^{-1} in aqueous mercurous nitrate solution was assigned as the Hg–Hg stretching mode of the Hg_2^{2+} ion. In this particular case the vibration would be formally inactive in the infrared and, in fact, many of the data on metal–metal stretching frequencies have been obtained by Raman spectroscopy. This is partly because the Raman lines arising from metal–metal stretching modes are often very intense (due in part to the large polarizability of the heavy metal atom involved) whereas in the infrared they are frequently weak or formally forbidden. For instance, the symmetric Sn–Sn stretching frequency in $(C_6H_5)_6Sn_2$ would be inactive in the infrared, although it would be formally allowed in $(C_6H_5)_3SnSn(C_2H_5)_3$. However, it has often been found (although not in this case) that the band cannot be detected or is very weak in the infrared. Even where binuclear complexes with *different* metals are involved, the mode often cannot be detected in the far infrared. For example, no infrared band was observed in the metal–metal stretching region of the complex $(CO)_5Mn–Re(CO)_5$ although an intense Raman line was found at 182 cm^{-1} (Gager *et al.* 1966).

A possible disadvantage of the Raman method is that such compounds are often very photosensitive and may be decomposed by the exciting radiation but use of lower-frequency lasers for excitation may overcome this problem where it occurs.

The list of observed bands assigned to "metal–metal stretching" is growing rapidly but it should be borne in mind that such vibrations will often be considerably mixed with other low-frequency vibrations in the normal mode and hence, in general, highly specific ranges for such modes can hardly be expected. Obviously, a fuller discussion of this topic must await more detailed normal coordinate calculations which are, as yet, virtually non-existent. One difficulty is that such metal–metal bonds are often part of complicated systems which are not readily amenable to calculation. If, for example, the metal–metal bond has adjacent bonds which have similar force constants and which link atoms of similar masses, the potential energy of the normal mode may be spread over several bonds fairly uniformly and it becomes meaningless to talk of a metal–metal stretch. In the case of the series $XHgMn(CO)_5$ (where X = Cl, Br and I) studied recently by Brier *et al.* (1967) a simple calculation showed that in the chloride and iodide it was reasonable to describe the bands at 184 cm^{-1} and 176 cm^{-1}, respectively, as the Hg–Mn stretches, whereas in the case of the bromide it was impossible to assign the bands at 195 cm^{-1} and 162 cm^{-1} to Hg–Br and Hg–Mn stretches due to the considerable mixing of these two motions in the normal mode.

At present, the greatest volume of metal–metal stretching frequency data relates to, (a) binuclear systems, and a smaller amount to (b) linear, and (c) bent, trinuclear systems. Polynuclear skeletons of metal–metal bonded systems or "metal cluster" compounds are considered in Section IV.

A. Binuclear Systems

Notwithstanding the comments above, it can be seen from the following tables that, provided the molecule does not have a centre of

TABLE XVII. Hg–Hg Stretching Frequencies in Binuclear Mercury Compounds. (Observed in the Raman.)

Compound	State	ν_{Hg-Hg} (cm^{-1})	Reference
$Hg_2(NO_3)_2$	aqueous solution	169	Woodward (1934)
$Hg_2(NO_3)_2.2H_2O$	solid	170	Gager *et al.* (1966)
		179.8	Stammreich and Sans (1967)
Hg_2Cl_2	solid	166	Poulet and Mathieu (1963), Gager *et al.* (1966) and Stammreich and Sans (1967)
Hg_2F_2	solid	185.9	Stammreich and Sans (1967)
Hg_2Br_2	solid	133	Gager *et al.* (1966) and Stammreich and Sans (1967)
Hg_2I_2	solid	112.5	Stammreich and Sans (1967)
$Hg_2(ClO_4)_2.4H_2O$	solid	181.6	Stammreich and Sans (1967)
$Hg_2(ClO_4)_2$	aqueous solution	173.3	Stammreich and Sans (1967)
Hg_2SO_4	solid	172.2	Stammreich and Sans (1967)
Hg_2CO_3	solid	174.5	Stammreich and Sans (1967)

TABLE XVIII. Hg–M Stretching Frequencies in Binuclear Hg–M Compounds.

Compound	State	M–Hg	ν_{M-Hg} (cm^{-1})	Raman/IR	Reference
$ClHgMn(CO)_5$	solid	Hg–Mn	184	IR	Brier *et al.* (1967)
$IHgMn(CO)_5$	solid	Hg–Mn	176	IR	Brier *et al.* (1967)
$Os(CO)_4(HgCl)Cl$	solid	Hg–Os	172, 155	Raman	Bradford *et al.* (1968)

TABLE XIX. Sn–Sn Stretching Frequencies in Binuclear Tin Compounds.

Compound	State	ν_{Sn-Sn} (cm^{-1})	IR/Raman	Reference
$(C_6H_5)_2Sn_2$	liquid	190	Raman	Brown *et al.* (1960)
$(C_6H_5)_2Sn_2$	solid	208	Raman	Gager *et al.* (1966)
$Ph_3SnSnEt_3$	liquid	208	IR	Carey and Clark (1967)
$Et_3SnSnBu^n_3$	liquid	199	IR	Carey and Clark (1967)
$Ph_3SnSnMe_3$	solid	194	IR	Carey and Clark (1967)
$Sn(SnPh_3)_4$	solid	207	Raman	Gager *et al.* (1966)

symmetry, the infrared as well as the Raman can be a fruitful source of information. Data of both types are included.

Some very approximate M–M stretching force constants for the carbonyl compounds of type $M_2(CO)_{10}$ have been calculated. The values k(Mn–Mn) = 1.4, k(Mn–Re) = 2.3 and k(Re–Re) = 1.6 md/Å are obviously not reliable as absolute values but may be significant as a trend, pointing to an unusual order of bond strengths (Gager *et al.* 1966).

Very recently, however, a thorough infrared and Raman study has been made on compounds of the type $Cl_3MCo(CO)_4$ where M = Si, Ge and Sn (Watters *et al.*, 1969). The $Cl_3SiCo(CO)_4$ is known to have a C_{3v} structure from X-ray work and the similarity of the spectra of the other two compounds suggests the same structure. Normal coordinate calculations using a modified valence force field yielded the following metal–metal stretching force constants: $f_{Co-Si} = 1.32$, $f_{Co-Ge} = 1.05$ and $f_{Co-Sn} = 1.23$ md/Å. They pointed out that the same force constants calculated on the basis of a pseudodiatomic model using one M–M assignment differ considerably from the values obtained by considering the whole molecule. In fact, not only the values differed, but also the *order* of the values thus illustrating that even trends based on such values must be treated with reserve.

TABLE XX. Sn–M Stretching Frequencies in Binuclear Sn–M Compounds.

Compound	State	Sn–M	ν_{Sn-M} (cm^{-1})	Raman/IR	Reference
$Ph_3SnMn(CO)_5$	solution	Sn–Mn	174	Raman	Gager *et al.* (1966)
$Ph_3SnMn(CO)_5$	solid	Sn–Mn	174	IR	Carey and Clark (1967, 1968)
$(CH_3)_3SnMn(CO)_5$	liquid	Sn–Mn	182	IR	Carey and Clark (1967, 1968)
$(CH_3)_2ISnMn(CO)_5$	solid	Sn–Mn	179	IR	Carey and Clark (1967, 1968)
$(CH_3)_2BrSnMn(CO)_5$	solid	Sn–Mn	191	IR	Carey and Clark (1967, 1968)
$(CH_3)_2ClSnMn(CO)_5$	solid	Sn–Mn	197	IR	Carey and Clark (1967, 1968)
$CH_3Cl_2SnMn(CO)_5$	solid	Sn–Mn	201	IR	Carey and Clark (1967, 1968)
$Cl_3SnMn(CO)_5$	solid	Sn–Mn	201	IR	Carey and Clark (1967, 1968)
$(CH_3)_3SnGe(C_6H_5)_3$	solid	Sn–Ge	225	IR	Carey and Clark (1967)
$(C_2H_5)_3SnGe(C_6H_5)_3$	solid	Sn–Ge	230	IR	Carey and Clark (1967)
$(C_6H_5)_3SnGe(C_4H_9)_3$	solid	Sn–Ge	235	IR	Carey and Clark (1967)
$(CH_3)_3SnMo(CO)_3C_5H_5$	solid	Sn–Mo	172	IR	Carey and Clark (1967, 1968)
$(C_6H_5)_3SnMo(CO)_3C_5H_5$	solid	Sn–Mo	169	IR	Carey and Clark (1967, 1968)
$Cl(CH_3)_2SnMo(CO)_3C_5H_5$	solid	Sn–Mo	197	IR	Carey and Clark (1968)
$Cl_2CH_3SnMo(CO)_3C_5H_5$	solid	Sn–Mo	195	IR	Carey and Clark (1968)
$Cl_3SnMo(CO)_3C_5H_5$	solid	Sn–Mo	190	IR	Carey and Clark (1968)
$(CH_3)_3SnW(CO)_3C_5H_5$	solid	Sn–W	169	IR	Carey and Clark (1968)
$Cl(CH_3)_2SnW(CO)_3C_5H_5$	solid	Sn–W	175	IR	Carey and Clark (1968)
$(CH_3)_3SnFe(CO)_2C_5H_5$	liquid	Sn–Fe	185	IR	Carey and Clark (1967, 1968)
$(C_6H_5)_3SnFe(CO)_2C_5H_5$	solid	Sn–Fe	174	IR	Carey and Clark (1967, 1968)
$(CH_3)_3SnCo(CO)_4$	solid	Sn–Co	176	IR	Carey and Clark (1967)

TABLE XXI. M–M′ Stretching Frequencies in Other Binuclear M–M′ Compounds.

Compound	State	M–M′	ν_{M-M} (cm^{-1})	Raman/IR	Reference
[a] Cd^{2+}	melt	Cd–Cd	183	Raman	Corbett (1962)
$Mn_2(CO)_{10}$	solution	Mn–Mn	157	Raman	Gager *et al.* (1966)
$MnRe(CO)_{10}$	solution	Mn–Re	182	Raman	Gager *et al.* (1966)
$Re_2(CO)_{10}$	solution	Re–Re	128	Raman	Cotton and Wing (1965) and Lewis *et al.* (1965)
$Ge_2(CH_3)_6$	liquid	Ge–Ge	273	Raman	Brown *et al.* (1960)

[a] Observed in a melt of $Cd_2(AlCl_4)_2$ and $Cd(AlCl_4)_2$.

B. Linear Trinuclear Species

In systems containing a linear M–M′–M metal skeleton, two vibrational modes would be expected, mainly associated with this part of the molecule. In centrosymmetric molecules, they correspond to a symmetric and antisymmetric stretch, the two being mutually exclusive in the infrared and Raman. This is well illustrated in the work of Hartley *et al.* (1968) who studied infrared and Raman spectra of $Os_3(CO)_{12}X_2$ (X = Cl, Br and I) which has a linear X–Os–Os–Os–X skeleton (D_{4h} symmetry for whole molecule). The whole heavy-atom skeleton should give rise to two Raman lines (a_{1g}) and two non-coincident infrared lines (a_{2u}), two being essentially metal-metal stretches and two being metal-X stretches. These bands are summarized in Table XXII.

TABLE XXII. M–X and M–M Stretching Frequencies (cm^{-1}) in Linear $Os_3(CO)_{12}X_2$ Systems (X = Cl, Br and I) in the Solid State.

Compound	ν_{Os-Os} Raman (a_{1g})	ν_{Os-Os} IR (a_{2u})	ν_{Os-X} Raman (a_{1g})	ν_{Os-X} IR (a_{2u})
$Os_3(CO)_{12}Cl_2$	116	162	313	304
$Os_3(CO)_{12}Br_2$	100	157	212	203
$Os_3(CO)_{12}I_2$	97	153	170	169

It is seen that the two higher-frequency modes are much more sensitive to change of X than the other pairs and can thus be assigned to essentially antisymmetric M–X stretching modes and the lower two as essentially M–M stretches. A normal coordinate analysis, however, indicates that the M–X and M–M stretching motions are considerably mixed, especially for the heavier halides. Values for stretching of the Os–Os bonds are about 1.6–1.7 md/Å based on a valence force field and making the further approximation of a linear pentatomic system.

Stretching frequencies for some other linear trinuclear species are given in Table XXIII.

TABLE XXIII. M–M Stretching Frequencies in Linear Trinuclear M–M′–M Compounds.

Compound	State	M–M′–M	$\nu_{M-M'}$ (cm^{-1})	Raman/IR	Reference
$[Co(CO)_4]_2Hg$	solution	Co–Hg–Co	161 (sym)	Raman	Stammreich *et al.* (1961)
			197 (a.s.)	IR	Brier *et al.* (1967)
$[Mn(CO)_5]_2Hg$	solid	Mn–Hg–Mn	186 (a.s.)	IR	Brier *et al.* (1967)
$[C_5H_5(CO)_3Mo]_2Hg$	solid	Mo–Hg–Mo	167 (a.s.)	IR	Brier *et al.* (1967)
$[C_5H_5(CO)_3W]_2Hg$	solid	W–Hg–W	169 (a.s.)	IR	Brier *et al.* (1967)
$[Co(CO)_4]_2Cd$	solution	Co–Cd–Co	152 (sym)	Raman	Stammreich *et al.* (1961)

TABLE XXIV. M–M′ Stretching Frequencies in Bent Trinuclear M⟋M′⟍M Compounds.

Compound	State	M⟋M′⟍M	$\nu_{M'-M}$ (cm^{-1}) Raman	IR	Reference
cis $Fe(CO)_4(HgCl)_2$	solid	Hg–Fe–Hg	201 186	204 186	Bradford *et al.* (1968)
$Fe(CO)_4(HgBr)_2$	solid	Hg–Fe–Hg	182 169	184 169	Bradford *et al.* (1968)
$Fe(CO)_4(HgI)_2$	solid	Hg–Fe–Hg	152	155	Bradford *et al.* (1968)
$Ru(CO)_4(HgCl)_2$	solid	Hg–Ru–Hg	185 —	187 170	Bradford *et al.* (1968)
$Ru(CO)_4(HgBr)_2$	solid	Hg–Ru–Hg	172 —	175 157	Bradford *et al.* (1968)
$Os(CO)_4(HgCl)_2$	solid	Hg–Os–Hg	— —	169 156	Bradford *et al.* (1968)
$Os(CO)_4(HgBr)_2$	solid	Hg–Os–Hg	154 —	156 140	Bradford *et al.* (1968)
$Os(CO)_4(HgI)_2$	solid	Hg–Os–Hg	— 135	140 128	Bradford *et al.* (1968)
$Et_4N^+[Rh_2Cl_2(SnCl_3)_4]$ [a]	solid	Sn–Rh–Sn	—	208.6	Adams and Chandler (1965)
$Et_4N^+[Rh_2Br_2(SnBr_3)_4]$ [a]	solid	Sn–Rh–Sn	—	217.1	Adams and Chandler (1965)

[a] Square planar, chlorine bridged complexes.

C. Bent Trinuclear Species

Compounds of the type $Fe(CO)_4(HgCl)_2$ have a *cis* configuration resulting in a bent Hg╲Fe╱Hg skeleton. Examples of this type of compound are listed in Tables XXIV and XXV.

TABLE XXV. Metal-Metal Stretching Frequencies in Miscellaneous Compounds.

Compound	State	M–M	ν(M–M) (cm^{-1})	Raman/IR	Reference
$Fe(CO)_4Hg$(polymeric)	solid	Fe–Hg	196	IR	Adams *et al.* (1967) Brier *et al.* (1967)
$Et_4N^+[Pt(SnCl_3)_5]$	solid	Sn–Pt	209.8	IR	Adams and Chandler (1965)
$Et_4N^+[Pt(SnBr_3)_5]$	solid	Sn–Pt	207.3 192.6	IR	Adams and Chandler (1965)

IV. Metal-Cluster Compounds

These are an interesting class of compounds in which the subjects of bridging halogens and metal–metal bonds become intimately mixed. They may be defined as compounds which contain a group of metal atoms bonded together entirely or partly by direct metal–metal bonds and which may also include non-metal atoms in bridging and terminal positions on the metal. This definition, together with a review of the preparation, properties and structures of such system has been given by Cotton (1966).

The literature on vibrational spectra of these systems is extremely sparse at present although it will undoubtedly prove to be a fruitful method of investigation. Many of the molecules are coloured and most of the published work to date is concerned with far-infrared spectra. However, in many of these systems in which highly symmetrical species are present, for example $[Mo_6Cl_8]^{4-}$, only a few infrared-active frequencies are expected (*vide infra*) and it is obvious that Raman spectroscopy is highly relevant.

Halide cluster compounds of Mo and W (Clark *et al.* 1966; Cotton *et al.* 1967) and Nb and Ta (Boorman and Straughan 1966; Mackay and Schneider 1968) have been investigated in the far-infrared recently. A more detailed infrared and Raman study of the $[(Mo_6Cl_8)X_6]^{2-}$ species has also been carried out (Hartley and Ware 1967). There has been a brief report on the Raman spectrum of $Ir_4(CO)_{12}$ by Quicksall and Spiro (1967) and Raman and far-infrared spectra of a series of $Os_3(CO)_{12}$ derivatives has recently been published by Hartley *et al.* (1968).

Considering the simplest series first, $Os_3(CO)_{12}$ is known from an X-ray study by Corey and Dahl (1962) to have the three osmium atoms located at the corners of an equilateral triangle. The point group for this is D_{3h} and two metal-metal stretching modes are expected on this basis, a_1' (Raman) and e' (Raman and infrared). The observed infrared and Raman spectra were consistent with the selection rules and the frequencies are listed in Table XXVI.

TABLE XXVI. Raman and IR-Active Metal-Metal Stretching Frequencies and MM Stretching Force Constant in $Os_3(CO)_{12}$ and $Ru_3(CO)_{12}$.

	$\nu_{(M-M)}$ (cm^{-1}) Raman	IR	k_{M-M} (md/Å)
$Ru_3(CO)_{12}$	186 (a_1')		
	152 (e')	n.i.	1.6
$Os_3(CO)_{12}$	161 (a_1')	—	
	120 (e')	121	1.7

n.i.—not investigated.

Frequencies in the Os system are less than the Ru system, as would be expected from the mass effect; the M–M stretching force constants (computed using a SVFF applied to the pseudotriatomic $M_3(CO)_{12}$ system) are also shown and are seen to be similar in the two molecules and of the same order as those found for other metal-metal stretches (Section III). The low-frequency spectra of $Os_3(CO)_{10}(OCH_3)_2$ were consistent with three Os atoms at the corners of an isosceles triangle (C_{2v}).

In $Ir_4(CO)_{12}$ the iridium atoms are known to be in a tetrahedral cluster (Wei and Dahl 1966). Considering only the iridium skeleton, three normal modes are expected of species a_1, t_2 and e, all three being active in the Raman. These were observed in the Raman by Quicksall and Spiro (1967) at 208, 164 and 105 cm^{-1}, respectively; using a SVFF applied to the Ir skeleton, a single Ir–Ir stretching force constant of 1.3 md/Å predicted all three of these frequencies to within 10 cm^{-1}.

The $[Mo_6Cl_8]^{4+}$ and $[W_6Cl_8]^{4+}$ systems have received some attention recently as the $[Mo_6X_8Y_6]^{2-}$ ions, where X and Y = Cl, Br and I (Clark *et al.* 1966; Cotton *et al.* 1967). The former authors studied the infrared spectra of these systems to 200 cm^{-1} and the latter to 80 cm^{-1}. These $[M_6X_8Y_6]^{2-}$ systems have O_h symmetry with the six metal atoms at the apices of an octahedron. The eight halogen atoms (X) are in bridging positions one above each face of the octahedron and the six halogen atoms (Y) are in terminal positions on the metal atoms, the MY bonds being directed outwards from the centre of the octahedron. For O_h symmetry, the vibrational representation for the system is:

$$3a_{1g} + 3e_g + 2t_{1g} + 4t_{2g} + a_{2u} + e_u + 5t_{1u} + 3t_{2u}.$$

The only infrared-active modes for this point group are those belonging to the t_{1u} species and, for the $[M_6X_8Y_6]^{2-}$ systems, there are five of these. The symmetry coordinate representation of these predicts that two will be M–X stretching modes, one an M–Y stretch and another a M–M stretch—although this would be an oversimplification. The proposed assignments in these systems are summarized in Table XXVII (Cotton *et al.* 1967).

TABLE XXVII. Far-Infrared Spectra (t_{1u}) for $[M_6X_8Y_6]^{2-}$ Systems.

Compound	t_{1u} modes (approx. description) cm^{-1} MX str.	M–Y str.	XMY def.	M–M str.
$[Mo_6Cl_8Cl_6]^{2-}$	350–310	246	110	220
$[Mo_6Cl_8Br_6]^{2-}$	330–290	168	77(?)	232
$[Mo_6Cl_8I_6]^{2-}$	325–280	132	70	233
$[Mo_6Br_8Cl_6]^{2-}$	345–330, 300	260	120	240–260(?)
$[Mo_6Br_8Br_6]^{2-}$	270, 250–220	130–170	80	220(?)
$[Mo_6Br_8I_6]^{2-}$	274, 250–210	120	80	210–250
$[W_6Cl_8Cl_6]^{2-}$	320–290	224	205(?)	150(?)

One of the criteria used in these assignments was that other evidence (see Cotton 1966) indicates that the M–Y type of bonds are longer than the M–X type. From a consideration of the $[Mo_6Cl_8X_6]^{2-}$ (X = Cl, Br and I) systems they concluded that the metal metal bonding in the cluster was not much affected by changes of Y.

Infrared (solid state) and Raman (solution) spectra of the systems $H_2[(Mo_6Cl_8)X_6]8H_2O$ (X = Cl, Br and I) have been measured by Hartley and Ware (1967). That complementary Raman measurements are valuable is obvious from a glance at the vibrational representation above; a_{1g}, e_g and t_{2g} modes are active in the Raman, thus giving ten permitted fundamentals, whereas only five of the modes (t_{1u}) are active in the infrared. (The remaining seven modes, $2t_{2g}$, a_{2u}, e_u and $3t_{2u}$ are inactive in both infrared and Raman.) The assignments are listed below in Table XXVIII. The fact that the number of Raman lines agrees well with the expectations for the $[(Mo_6Cl_8)X_6]^2$ unit suggests that the Mo–X (that is, terminal) bonds have appreciable covalent character. Using a simple, 3-parameter force field comprising f_M (adjacent Mo atoms), f_C (adjacent Mo and bridging Cl atoms) and f_X (Mo–X bonds), they obtained a good fit between calculated and observed values for the active fundamentals. Force constants obtained on this basis were $f_M = 1.6$, 1.6 and 1.4 md/Å and $f_X = 1.8$, 1.6 and 1.3 md/Å, respectively for X = Cl, Br and I. The normal coordinate analysis confirmed that it is virtually meaningless to talk in terms of Mo–Mo and MoX stretching modes as these motions are considerably mixed in the normal coordinates.

TABLE XXVIII. Raman and Infrared Frequencies (cm^{-1}) of $[Mo_6Cl_8X_6]^{2-}$ Species (Hartley and Ware 1967).

Raman	Cl	Br	I	Assignment
(In methanol solution)	92 s (dp)	60 ms (dp)	44 s	t_{2g}
	179 ms (dp)	134 m (dp)	109 ms	e_g
	195 w (dp)	195 w	184 wm	t_{2g}
	215 w (dp)	216 m (dp)	215 mw	e_g
	225 w		150 w	t_{2g}
	236 s (pol)	160 s (pol)	117 s	a_{1g}
	247 s (dp)	249 s (dp)	247 m	t_{2g}
	310 (dp)	279 w (dp)	247	e_g
	318 vs (pol)	318 vs (pol)	318 s	a_{1g}
	366	356 w	338 mw	a_{1g}
		369 (vw)		
Infrared	221 m	167 s	132 s	t_{1u}
	247 ms	232 m	229 m	t_{1u}
	330 s (bd)	305 s (bd)	295 m	t_{1u}
	351 m	360 m (bd)	257 w	t_{1u}
	232 w (sh)	226 w (sh)	325 vw (sh)	other bands
	294 w (sh)			other bands

w = weak; m = medium; s = strong; sh = shoulder; bd = broad, dp = depolarized; pol = polarized.

Far-infrared spectra of M_6X_{14} (M = Nb and Ta; X = Cl and Br) have been investigated as anhydrous solids, hydrates, pyridine adducts and in solution by Boorman and Straughan (1966). Spectra of hydrated $M_6X_{12}(BF_4)_2$ and $M_6X_{12}(SO_4)$ were also examined. The spectra of all the systems contain four strong bands, characteristic of the $[Nb_6Cl_{12}]^{2+}$ species, and a similar pattern was observed for the $[Nb_6Br_{12}]^{2+}$ ions. Ideally, an isolated $[M_6X_{12}]^{2+}$ unit would have O_h symmetry and only four strong infrared-active (t_{1u}) modes are expected. Again, insofar as it is meaningful to separate these, three should be mainly associated with motions of the M–X bonds and the other with the M–M motion in the octahedron. Typical values for frequencies assigned as t_{1u} fundamentals are listed below for the $M_6X_{14}.8H_2O$ compounds.

	t_{1u} modes (cm^{-1})			
$Nb_6Cl_{14}.8H_2O$	343	280	244	145
$Nb_6Br_{14}.8H_2O$	265	206	172	140
$Ta_6Cl_{14}.8H_2O$	326	232	189	142
$Ta_6Br_{14}.8H_2O$	230	180	142	100

The complexity of the spectra increased on passing from M_6X_{14} in solution to $M_6X_{14}.8H_2O$ to M_6X_{14} (anhydrous), this being interpreted in terms of deviation from the octahedral symmetry of the isolated $[M_6X_{12}]^{2+}$ unit.

Complexes of the type $[Et_4N^+]_n[(Nb_6Cl_{12})Cl_6]$, which contain the $[Nb_6Cl_{12}]^{n+}$ ($n = 2$, 3 and 4) ions have been investigated in the infrared together with some other systems involving the $(Nb_6Cl_{12})^{2+}$ cluster (Mackay and Schneider 1968). The low-frequency assignments for the t_{1u} modes of these $(Nb_6Cl_{12})^{n+}$ clusters are summarized together with the corresponding frequencies for the $(Ta_6Cl_{12})^{3+}$ system below.

	$(Nb_6Cl_{12})^{2+}$	$(Nb_6Cl_{12})^{3+}$	$(Nb_6Cl_{12})^{4+}$	$(Ta_6Cl_{12})^{3+}$
ν_{M-Cl} (cm^{-1})	340	346	353	327
	279	290	302	285
	234	251	267	244
ν_{M-M} (cm^{-1})	143	142	141	140

The three higher frequencies in each species are assigned to vibrations mainly involved with MCl motions, and the band at about 140 cm^{-1} to motion mainly involved with the metal atom skeleton. The effect of the increasing positive charge (that is, increasing oxidation state) is less than is usually found in simpler systems (c.f. Section II2), but this is reasonable as the change in oxidation state for each metal atom is only one sixth of the change in the oxidation state of the cluster.

V. Structural Diagnosis

In principle, at least, inference of structure can be made from the observation of infrared and Raman spectra without recourse to a detailed assignment of vibrational frequencies; this is one of the most important and fruitful applications of vibrational spectroscopy.

In outline, the following procedure is used.

(1) Prediction of the various possible geometries for a given stoichiometry and, from a consideration of the symmetry elements possessed by each, deduction of the point group to which each geometry belongs.

(2) From the character table for each point group and by counting the number of atoms unshifted by each of the symmetry operations, the vibrational representation (that is, the distribution of the 3N-6 normal modes among the various symmetry species) is deduced.

(3) From this distribution several important deductions are made: (i) the degeneracy (if any) of each species, (ii) the activity of each species (that is, infrared, Raman, or inactive), and (iii) the number of totally symmetric modes. These procedures are fully described in "Molecular Vibrations" by Wilson, Decius and Cross (Wilson *et al.* 1955) and conveniently summarized by Drago (1965) and Ware (1968). These deductions provide a number of useful criteria which may be used to decide between various possible structures. Firstly, (i) and (ii) predict the number of fundamental frequencies active in the infrared and in the Raman, and the number of these that should be coincident in the two types of spectra. A special case of this is the *rule of mutual exclusion*; that is, if the molecule has a centre of symmetry there will be no coincident Raman and infrared bands. Secondly, (iii) predicts the number of polarized Raman lines as these only occur for totally symmetric modes.

These criteria are often sufficient to decide between various possible structures. If the molecule can be examined in the gaseous state, additional information can be obtained from the contours of the vibration-rotation bands in the infrared, although this is obviously of limited application in inorganic chemistry.

This procedure is not always so straightforward in practice. For example, Raman polarization data can often only be obtained on samples in the liquid state or in solution; this is not always possible. Some allowed bands may be inherently weak and be undetected; the presence of overtone and combination bands may complicate the spectrum and if, for any reason, (see Section I) there is a relaxation of the selection rules in the solid state, this may lead to the appearance of formally forbidden bands. This in turn may lead to false conclusions from, for example, the number of infrared and Raman coincidences.

Three examples below illustrate the usefulness of the method. In the first example, the structure of $[As^{III}Cl_4]^-$, a choice is required between various possible stereochemistries. In the second, P_2I_4, the problem is to decide between the various possible conformations which could arise due to rotation about the P–P bond. In the case of the $[MX_4Y_2]^{2-}$ systems, positional isomerism is possible and a choice between the *cis* and *trans* isomers is required.

The $[As^VCl_4]^+$ ion is tetrahedral and the vibration spectrum is consistent with this structure (Weidlein and Dehnicke 1965). However, in a study of the $[As^{III}Cl_4]^-$ ion, Davies and Long (1968) found that too many strong lines were obtained for a tetrahedral structure. Four possible geometries for such an MX_4^- entity could be envisaged:

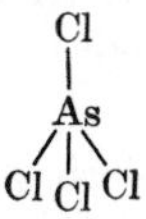

T_d

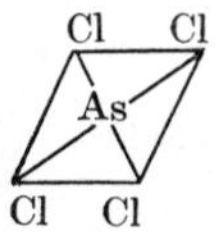

D_{4h}

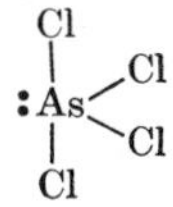

C_{2v}

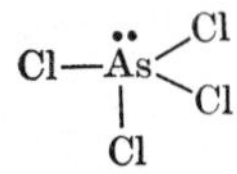

C_{3v}

The expectations for numbers of modes, activities, and so on, for these possibilities are shown in Table XXIX.

TABLE XXIX.

Point Group	No. of fundamental frequencies	No. active in IR	No. active in Raman	No. of IR/Raman coincidences	No. of polarized Raman bands
T_d	4	2	4	2	1
D_{4h}	7	3	3	0	1
C_{2v}	9	8	9	8	4
C_{3v}	6	6	6	6	3

Eight Raman lines were found of which seven were considered to be fundamentals. Of these three were polarized, two depolarized and the remaining two lines were too weak to determine their state of polarization. The far-infrared spectrum consisted of eight lines, seven of which were considered to be fundamentals and there were at least six coincidences between infrared and Raman bands. Thus, without any assignments being necessary, these data rule out the structures of point groups T_d, D_{4h} and C_{3v} and suggest the C_{2v} structure.

P_2I_4 has been reported to have a *trans* C_{2h} structure in the solid state but a gauche structure when dissolved in CS_2. Frankiss *et al.* (1967) have studied low-frequency Raman and infrared spectra in both solid and solution and found the two to be closely similar. The choice of configurations is between a *cis* C_{2v} structure (that is, the two PI_2 groups in an eclipsed position), a *trans* C_{2h} structure (with a centre of symmetry) and a *gauche* C_2 form which is intermediate between the *cis* and *trans* forms. The expectations for these conformations are summarized in Table XXX.

TABLE XXX.

Confor-mation	Point group	No. of Raman-active fundamentals	No. IR-active fundamentals	No. IR/Raman coincidences	No. of Raman-active stretching modes	
					Polarized	Depolarized
Trans	C_{2h}	6	6	0	2	1
Gauche	C_2	12	12	12	3	2
Cis	C_{2v}	12	9	9	2	3

The P_2I_4 solution spectra showed no coincidences between infrared and Raman bands (within the experimental error), thus suggesting a *trans* structure. The observation of only six Raman-active and four infrared-active fundamentals also favours the *trans* structure. Also, two Raman-active stretching fundamentals are polarized and one depolarized which is consistent with the *trans* configuration. The similarity between solution and solid-state spectra confirms that the solid is in the *trans* configuration, consistent with the X-ray results.

Mixed octahedral haloanions of the type $[MX_4Y_2]^{n-}$ and $[MX_3Y_3]^{n-}$ have various possibilities for positional isomerism (see Table XXXIII). A number of mixed halogen complexes of Ti^{IV} and Sn^{IV} of the $[MX_4Y_2]^{2-}$ type have recently been investigated by Clark *et al.* (1968b). The choice here is between a C_{2v} (*cis*) and D_{4h} (*trans*) configuration. These samples were examined as powders and, hence, no Raman polarization data could be expected. For the C_{2v} structure, fifteen lines would be expected, all of which should be Raman-active and all except the two a_2 modes infrared-active (i.e. thirteen coincidences). For D_{4h}, there should be eleven lines (five infrared-active and six Raman-active) with no coincidences due to the centre of symmetry. From the pattern of bands obtained in infrared and Raman spectra in the SnCl and SnBr stretching regions (which are known approximately from studies in $[SnCl_6]^{2-}$ and $[SnBr_6]^{2-}$ systems) it is apparent that, for $[SnCl_4Br_2]^{2-}$, $[SnCl_4I_2]^{2-}$ and $[SnBr_4Cl_2]^{2-}$ at least, the complexes are in the *cis* configuration. For example, the infrared and Raman lines observed for $[SnCl_4Br_2]^{2-}$ are as follows.

	ν(SnCl) (cm^{-1})	ν(SnBr) (cm^{-1})
Raman	204, 287, 264, 254	204, 180
Infrared	306, 290, 267, 252	205, 184

From Table XXXIII it is seen that the numbers of stretching frequencies agree well with the expectations for the C_{2v} symmetry. This, together with the coincidences between infrared and Raman spectra, strongly suggests that the C_{2v} structure obtains. The same configuration probably occurs in the corresponding $[TiX_4Y_2]^{2-}$ species, although the experimental evidence is less clear cut.

As mentioned earlier, useful structural information can be obtained from far-infrared spectra alone, and many examples of successes in this field could be quoted, although evidence of this type is not usually so decisive as that obtained from a complementary study with Raman spectra. Apart from relying on fewer criteria for the choice of structure, this method relies to some extent upon the concept of group vibrations and, hence, the results must be treated with some caution.

The vibrational representations for a number of commonly occurring MX_n (M = metal, X = halogen) and mixed MX_nY_m species are listed in Tables XXXI to XXXIV together with the expected number of polarized Raman lines and number of infrared and Raman coincidences. In the mixed systems, modes mainly associated with MX or MY stretching are indicated, although MX and MY vibrations of the same symmetry may be considerably mixed together in a normal mode, depending on the various masses and force constants. Thus, for example, it can be seen from Table XXXIII for a MX_4Y_2 species the expectation for a *trans* (D_{4h}) arrangement would be one infrared-active MX stretch (e_u), whereas a *cis* (C_{2v}) arrangement would give the possibility of four MX stretches ($2a_1 + b_1 + b_2$). Thus, in the MX stretching region (if that can be specified with reasonable certainty) the spectrum of the *cis* complex should have more bands than the *trans* complex. A number of hazards are involved with this procedure, for example failure to observe inherently weak bands, accidental degeneracy, intensity borrowing by weak overtone or combination tones from nearby fundamentals, and the lifting of degeneracy by site symmetry effects (Section IC1). Examples of this type of application are complexes of the type $SnCl_4.2L$ (for example, Beattie and Rule 1964; Beattie *et al.* 1963). The complexes $SnCl_4.2CH_3CN$ and $SnCl_4.2(CH_3)_2CO$ were consistent with a *cis* configuration, and $SnCl_4.2(CH_2)_4O$ and $SnCl_4.2Et_2S$ with a *trans* configuration from the pattern of observed frequencies in the SnCl stretching region.

TABLE XXXI. Vibrational Representations for Some 4-Coordinate Species.

Ion or molecule	Geometry	Point group	Number of fundamental frequencies	Distribution of modes: $\nu(MX)$	Distribution of modes: $\nu(MY)$	Distribution of modes: Deformations	Number of modes active (species): I.R.	Number of modes active (species): Raman	Number of modes active (species): Inactive	No. of IR/Raman coincidences	Number of polarized Raman lines
MX_4	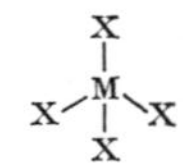	T_d	4	a_1+t_2	—	$e+t_2$	2 (t_2)	4 (a_1, e, t_2)	0	2	1
MX_4	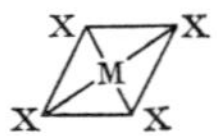	[a] D_{4h}	7	$a_{1g}+b_{1g}+e_u$	—	$a_{2u}+b_{2g}+b_{2u}+e_u$	3 (a_{2u}, e_u)	3 (a_{1g}, b_{1g}, b_{2g})	$1(b_{2u})$	0	1
MX_3Y	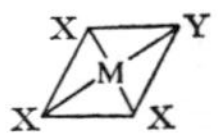	[b] C_{2v}	9	$2a_1+b_2$	a_1	$a_1+2b_1+2b_2$	9 (a_1, b_1, b_2)	9 (a_1, b_1, b_2)	0	9	4
MX_2Y_2 (*trans*)		[c] D_{2h}	9	a_g+b_{3u}	a_g+b_{2u}	$b_{1g}+2b_{1u}+b_{2u}+b_{3u}$	6 (b_{1u}, b_{2u}, b_{3u})	3 (a_g, b_{1g})	0	0	2
MX_2Y_2 (*cis*)	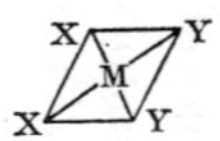	[b] C_{2v}	9	a_1+b_2	a_1+b_2	$2a_1+a_2+b_1+b_2$	8 (a_1, b_1, b_2)	9 (a_1, a_2, b_1, b_2)	0	8	4

[a] After Nakamoto. [b] z axis along C_2 axis; x axis perpendicular to plane. [c] z axis perpendicular to plane; x axis along XMX.

TABLE XXXII. Vibrational Representations for Some Planar Bridged Species.

Ion or molecule	Geometry	Point group	Number of fundamental frequencies	Distribution of modes: $\nu(MX)$ Terminal	$\nu(MX)$ Bridging	$\nu(MY)$	Deformations	No. of modes active (Species): I.R.	Raman	Inactive	No. of IR/Raman coinci-dences	No. of polarized Raman lines
M_2X_6	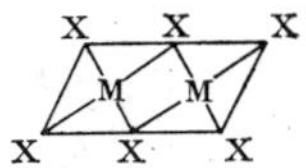	[a] D_{2h}	18	$a_g+b_{1g}+b_{2u}+b_{3u}$	$a_g+b_{1g}+b_{2u}+b_{3u}$	—	$2a_g+b_{1g}+b_{2g}+b_{3g}+a_u+2b_{1u}+b_{2u}+b_{3u}$	8 (b_{1u}, b_{2u}, b_{3u})	9 $(a_g, b_{1g}, b_{2g}, b_{3g})$	1 (a_u)	0	4
$M_2X_4Y_2$ *trans*	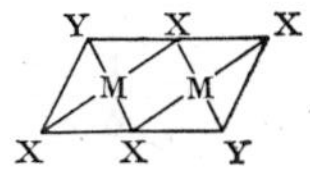	[b] C_{2h}	18	a_g+b_u	$2a_g+2b_u$	a_g+b_u	$3a_g+3a_u+2b_g+2b_u$	9 (a_u, b_u)	9 (a_g, b_g)	0	0	7
$M_2X_4Y_2$ *cis*	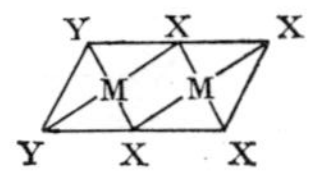	[c] C_{2v}	18	a_1+b_2	$2a_1+2b_2$	a_1+b_2	$3a_1+2a_2+3b_1+2b_2$	16 (a_1, b_1, b_2)	18 (a_1, a_2, b_1, b_2)	0	16	7
$M_2X_2Y_4$	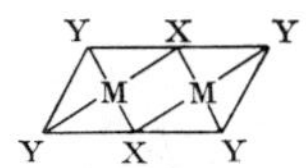	[a] D_{2h}	18	—	$a_g+b_{1g}+b_{2u}+b_{3u}$	$a_g+b_{1g}+b_{2u}+b_{3u}$	$2a_g+b_{1g}+b_{2g}+b_{3g}+a_u+2b_{1u}+b_{2u}+b_{3u}$	8 (b_{1u}, b_{2u}, b_{3u})	9 $(a_g, b_{1g}, b_{2g}, b_{3g})$	1 (a_u)	0	4

[a] z axis perpendicular to plane, y axis along MM line. [b] z axis is C_2 axis; x axis along MM line. [c] z axis along C_2 axis; x axis perpendicular to plane.

TABLE XXXIII. Vibrational Representations for Some 6-Coordinate Species.

Ion or molecule	Geometry	Point group	Number of fundamental frequencies	Distribution of Modes (MX)	Distribution of Modes (MY)	Distribution of Modes Deformations	Number of modes active (species) I.R.	Number of modes active (species) Raman	Number of modes active (species) Inactive	No. of IR/Raman coincidences	No. of polarized Raman lines
MX_6	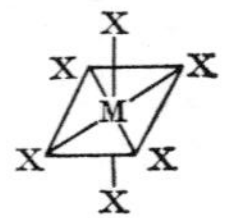	O_h	6	$a_{1g}+e_g+t_{1u}$	—	$t_{1u}+t_{2g}+t_{2u}$	$2(t_{1u})$	$3(a_{1g}, e_g, t_{2g})$	$1(t_{2u})$	0	1
MX_5Y	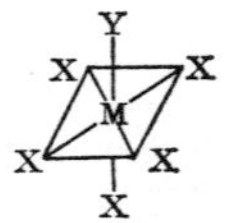	C_{4v}	11	$2a_1+b_1+e$	a_1	$a_1+b_1+b_2+3e$	8 (a_1, e)	11 (a_1, b_1, b_2, e)	0	8	4
MX_4Y_2 (*cis*)	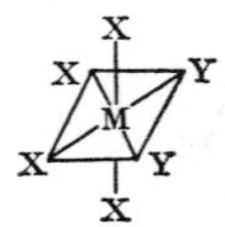	[a]C_{2v}	15	$2a_1+b_1+b_2$	a_1+b_1	$3a_1+2a_2+2b_1+2b_2$	13 (a_1, b_1, b_2)	15 (a_1, a_2, b_1, b_2)	0	13	6
MX_4Y_2 (*trans*)	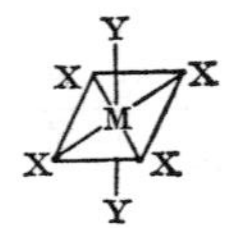	D_{4h}	11	$a_{1g}+b_{1g}+e_u$	$a_{1g}+a_{2u}$	$b_{2g}+e_g+a_{2u}+b_{2u}+2e_u$	5 (e_u, a_{2u})	6 $(a_{1g}, b_{1g}, b_{2g}, b_{2u}, e_g)$	0	0	2
MX_3Y_3 (*cis*)	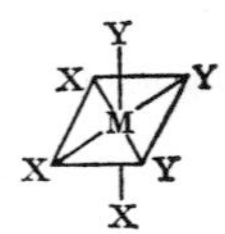	C_{3v}	10	a_1+e	a_1+e	$2a_1+a_2+3e$	9 (a_1, e)	9 (a_1, e)	$1(a_2)$	9	4
MX_3Y_3 (*trans*)	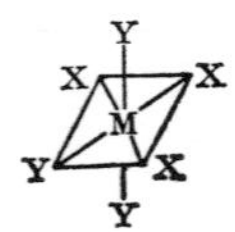	[b]C_{2v}	15	$2a_1+b_2$	$2a_1+b_1$	$2a_1+a_2+3b_1+3b_2$	14 (a_1, b_1, b_2)	15 (a_1, a_2, b_1, b_2)	0	14	6

[a] z axis along C_2 axis; x axis along XMX line. [b] z axis along C_2 axis; x axis along YMY line.

TABLE XXXIV. Vibrational Representations for Some 5-Coordinate Species.

Ion or molecule	Geometry	Point group	Number of fundamental frequencies	Distribution of modes ν(MX)	ν(MY)	Deformations	No. of modes active (species) IR	Raman	Inactive	No. of IR/Raman coincidences	No. of polarized Raman lines
MX_5	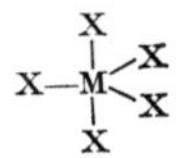	D_{3h}	8	$2a_1'+a_2''+e'$	—	$a_2''+2e'+e''$	5 (a_2'', e')	6 (a_1', e', e'')	0	3	2
MX_4Y	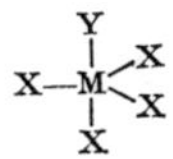	C_{3v}	8	$2a_1+e$	a_1	a_1+3e	8 (a_1, e)	8 (a_1, e)	0	8	4
MX_4Y	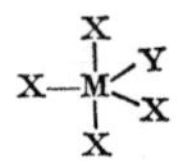	[a] C_{2v}	12	$2a_1+b_1+b_2$	a_1	$2a_1+a_2+2b_1+2b_2$	11 (a_1, b_1, b_2)	12 (a_1, a_2, b_1, b_2)	0	11	5
MX_3Y_2	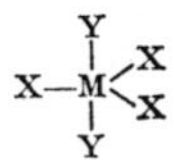	D_{3h}	8	$a_1'+e'$	$a_1'+a_2''$	$a_2''+2e'+e''$	5 (a_2'', e')	6 (a_1', e', e'')	0	3	2
MX_3Y_2	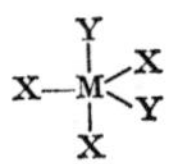	C_s	12	$2a'+a''$	$2a'$	$4a'+3a''$	12 (a', a'')	12 (a', a'')	0	12	8
MX_3Y_2		[a] C_{2v}	12	$2a_1+b_1$	a_1+b_2	$2a_1+a_2+2b_1+2b_2$	11 (a_1, b_1, b_2)	12 (a_1, a_2, b_1, b_2)	0	11	5
MX_5	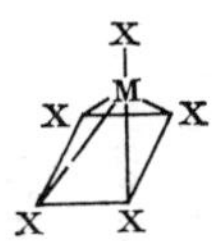	C_{4v}	9	$2a_1+b_1+e$	—	$a_1+b_1+b_2+2e$	6 (a_1, e)	9 (a_1, b_1, b_2, e)	0	6	3

[a] z axis along C_2 axis; x axis along XMX.

For trivalent complexes of the type $RhCl_3$.3-pyridine and $IrCl_3$. 3-pyridine, *cis*(C_{3v}) and *trans*(C_{2v}) isomers are possible (see Table XXXIII) and these species have been studied by Clark and Williams (1965) down to 200 cm^{-1}. From Table XXXIII it is seen that for the *trans* complex there are three infrared-active MCl stretches ($2a_1$ and b_2) and three infrared-active M–py stretches ($2a_1 + b_1$); in the *cis* complexes, two infrared MCl stretches ($a_1 + e$) and two M–py stretches ($a_1 + e$). In this example the stereochemistry was already known, but the observed spectra fitted the expectations well, indicating that the method is practical for structural inference.

VI. Complex Oxide Systems—Silicates and Borates

A. Introduction

Inorganic compounds of the general composition $X_mY_nO_z$ can be broadly divided into two general classes. The first class consists of complex oxides in which Y is an electropositive element, resulting in bonding forces that are predominantly ionic throughout the crystal structure. The second class consists of the salts of inorganic acids, which are characterized by the existence of discrete complex anions bound to the cations by ionic bonds but bound within themselves by covalent bonds. These complex anions are generally simple in form; although occasionally polynuclear, for example thionates, they are always finite in extent.

If, however, the Y atom in the general formula is neither strongly electropositive nor strongly electronegative (for example B, Si, P and Ge) a structural complexity arises due to the fact that anions of many different forms can occur. This is not to say that simple anions of these elements do not occur; BO_3^{3-} and SiO_4^{4-} ions are found in borates and silicates just as NO_3^- and SO_4^{2-} ions are found in nitrates and sulphates. The important factor is that in borates and silicates not only are the BO_3^{3-} and SiO_4^{4-} ions present but also many others of a much greater complexity are of common occurrence. It is for this reason that the anionic frameworks of these two systems are investigated. Concise and meaningful interpretation of the infrared data on borates and silicates must be based upon structural and chemical characteristics of the particular anionic framework being investigated.

B. The Silicates

The basic structural unit of the silicates is a silicon atom in four-fold coordination with oxygen (SiO_4). The bonds between the oxygen and silicon are such that four oxygens are always found at the corners of a tetrahedron of nearly constant dimensions and regular shape. This arrangement is consistent for all silicate compounds. Linkages of these tetrahedra in various compounds allow the silicates to be differentiated into five distinct structural classes, each possessing a symmetry and chemical character of basic importance to the interpretation of their infrared spectra (Berry and Mason 1959; Evans 1964).

In the literature, the nomenclature for the structural classes varies; to overcome this, a strictly descriptive nomenclature will be used to identify the various types of silicate networks.

1. *Independent Tetrahedra*

The silicon-oxygen tetrahedra are present as separate entities. The chemical formula is $(SiO_4)^{4-}$ and the suggested symmetry is T_d.

2. *Finite Linkage*

(a) *The Double Tetrahedra*

Two adjacent tetrahedra share one oxygen atom. The chemical formula is $(Si_2O_7)^{6-}$ and the suggested symmetry is C_{2v}.

(b) *The Ring Structure*

Two oxygens of each tetrahedron are shared with the neighbouring tetrahedra and the angular positions of the tetrahedra are such that closed units of a ring-like structure result. The general chemical formula is $(SiO_3)_n^{2-}$, where n signifies rings of $3(D_{3h})$, 4, and $6(D_{6h})$ silicons.

3. *Infinite Linkage*

(a) *Single Chain*

Tetrahedra are joined by the sharing of two oxygen atoms to produce a chain of infinite length. The general chemical formula is $[SiO_3^{2-}]\infty$.

(b) *Double Chain*

Tetrahedra are joined by alternately sharing two and three oxygens to form a chain of infinite length. The general chemical formula is $[(Si_4O_{11})^{6-}]\infty$ and the suggested symmetry is C_{2v}.

4. *Layer or Sheet Structure*

Three oxygens of each tetrahedron are shared with adjacent tetrahedra to form an extended flat sheet (a double chain structure in two directions of extension $(x - y)$ instead of one direction (x)). The general chemical formula is $(Si_2O_5)_n^{2-}$, where $n = 3$; the suggested symmetry is D_{3h}.

5. *Three-dimensional Network*

Each tetrahedron shares all of its oxygen atoms with other adjacent tetrahedra resulting in a three-dimensional network. The chemical formula is SiO_2.

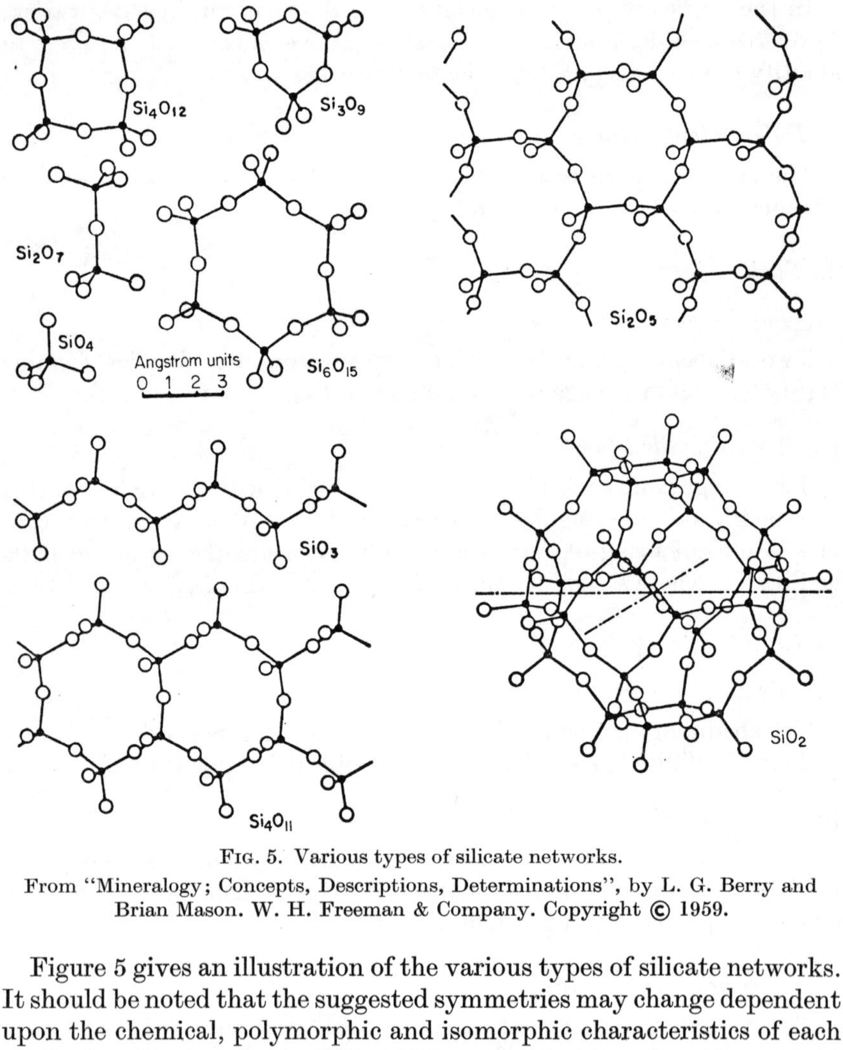

FIG. 5. Various types of silicate networks.
From "Mineralogy; Concepts, Descriptions, Determinations", by L. G. Berry and Brian Mason. W. H. Freeman & Company. Copyright © 1959.

Figure 5 gives an illustration of the various types of silicate networks. It should be noted that the suggested symmetries may change dependent upon the chemical, polymorphic and isomorphic characteristics of each particular silicate sample. This will become obvious when the work of Ishii *et al.* (1967) is discussed.

In the analysis of silicates, three basic precepts must be known and incorporated into the final interpretation of the infrared spectra of any silicate. (1) A knowledge of the chemical composition of the particular silicate is necessary. Generally, the silicate functions only as a network, orienting various cationic species. Frequencies in the far infrared are dependent on these M–O bonds and generally will vary as the cation changes. (2) The particular silicate network under consideration should be confirmed by X-ray analysis. Polymorphism as well as distortion of the silicate lattice is frequent in silicate minerals. Distortion of the silicate lattice changes the suggested symmetry and can cause the appearance (infrared active) or disappearance (infrared inactive) of bands. (3) Isomorphism is also of key importance in the far-infrared analysis of the silicates. The important factor to consider in this case is the similarity in size relations of the different ions rather than any chemical similarity. For example, aluminum and germanium, stable in both four-fold and six-fold coordination, can replace the silicon in the SiO_4 tetrahedra. When aluminum substitution occurs, there is a splitting or broadening of the infrared absorption bands. At times aluminum may also replace some of the six coordinated cations in the structure (Mg^{2+}, Fe^{2+}, Fe^{3+}, and so on). Thus it is necessary to have prior knowledge of the particular sample before an interpretation of its infrared spectra is attempted.

It has been found that the normal coordinate treatment of the silicates can best compensate for all these factors by the physical description of the vibrational modes. Ishii and co-workers (1967), by normal coordinate treatment on the ideal structure of dioctahedral- and trioctahedral-type micas, were able to assign absorptions due to the Si_2O_5 layer, the octahedral layer, and the interlayer vibrations in naturally occurring micas such as kaolinite, pyrophylite, talc, muscovite and others. Their study is of particular importance in that they consider each of the above precepts and account for any anomalies from the ideal structure of the micas.

Of first consideration is the structure of the layer silicates (Fig. 6). The structure consists of octahedral and tetrahedral layers. In an octahedral layer the sites of octahedral coordination are occupied by either tri- or divalent positive ions; generally Al^{3+} (dioctahedral) or Mg^{2+} (trioctahedral). The tetrahedral layer is a two-dimensional Si_2O_5 network. The entire structure of the layer silicates is determined by the stacking arrangement of these composite layers. Typical structural types are one-layer monoclinic (1M type), two-layer monoclinic (2M type) and three-layer trigonal (3T type). The micas, specifically muscovite and phlogopite, generally crystallize in 1M or 2M type structure.

	Dioctahedral Type	Trioctahedral Type
Octahedral Layer	Hydrargillite Type $Al(OH)_3$	Brucite Type $Mg(OH)_2$
1:1 Type Structure	Kaolinite $Al_2(Si_2O_5)(OH)_4$	
2:1 Type Structure	Pyrophyllite $Al_2(Si_2O_5)_2(OH)_2$	Talc $Mg_3(Si_2O_5)_2(OH)_2$
Mica	Muscovite $KAl_2(AlSi_3O_{10})(OH)_2$	Phlogopite $KMg_3(AlSi_3O_{10})(OH)_2$ Biotite Lepidomelane

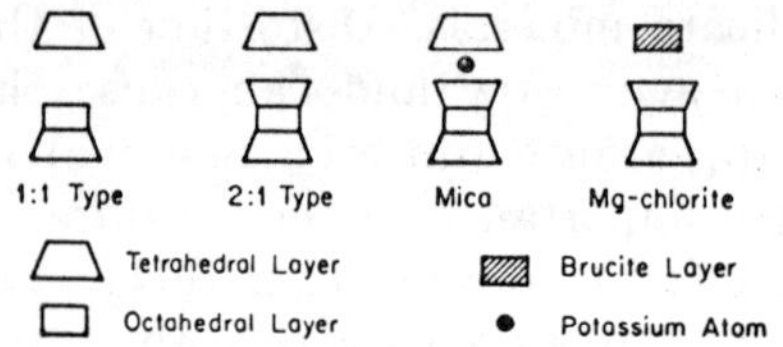

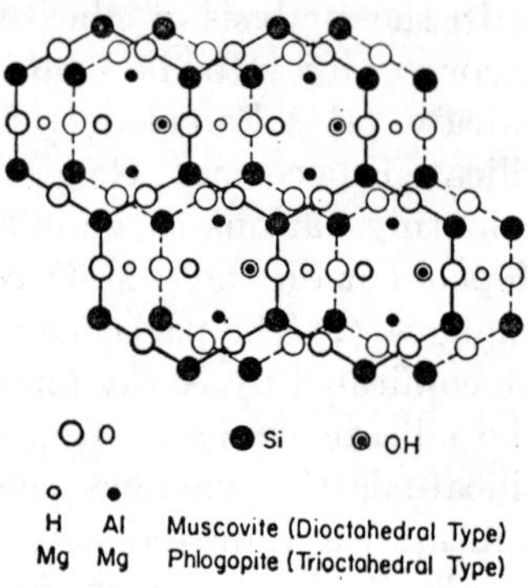

Fig. 6. Structure of layer silicates. ○ O, ● Si, ⊙ OH, ○ H, • Al Muscovite (Dioctahedral type), **O** Mg, . Mg (Trioctahedral type).

Table XXXV. Symmetry Species and Selection Rules for Si_2O_5 and $AlSi_3O_{10}$ Layers.

(1) Si_2O_5 layer

C_{6v}	N	T	n	IR	R	C_{3v}	N	T	n	IR	R
a_1	3	1	2	a	a	a_1	4	1	3	a	a
a_2	0	0	0	ia	ia						
b_1	3	0	5	ia	ia	a_2	3	0	3	ia	ia
b_2	1	0	1	ia	ia						
e_1	4	1	3	a	a	e	7	1	6	a	a
e_2	3	0	3	ia	a						

(Correlation: a_1, b_1 → a_1; a_2, b_2 → a_2; e_1, e_2 → e.)

(2) $AlSi_3O_{10}$ layer

C_s	N	T	n	IR	R
a′	26	2	24	a	a
a″	16	1	15	a	a

(3) Idealized 1M type mica

Dioctahedral type						Trioctahedral type					
C_{1h}^3	N	T	n	IR	R	C_{2h}^3	N	T	n	IR	R
a_g	16	0	16	ia	a	a_g	16	0	16	ia	a
b_g	14	0	14	ia	a	b_g	14	0	14	ia	a
a_u	15	1	14	a	ia	a_u	16	1	15	a	ia
b_u	21	2	19	a	ia	b_u	23	2	21	a	ia

IR infrared; R Raman; a active; ia inactive; N number of degrees of freedom; T number of overall translations of Bravais unit cell; n number of intramolecular vibrational modes.

Of critical importance for micas is isomorphism. Some of the Si atoms of the tetrahedral layer are substituted by Al atoms. The result is that the composite layer is negatively charged and cations (generally K^+ ions) are added simultaneously between the layers to keep the

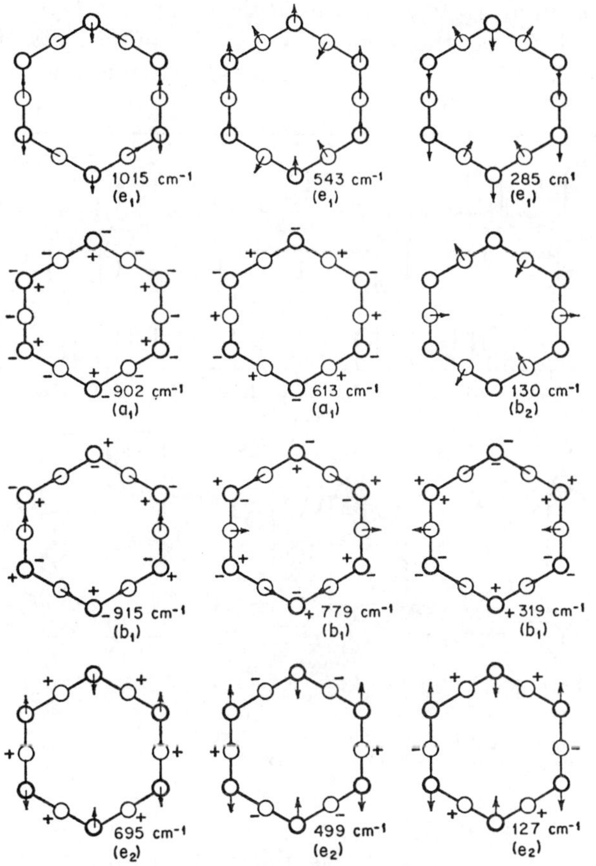

FIG. 7. Vibrational modes of the Si_2O_5 layer.

electrical neutrality. In muscovite and phlogopite, the substitution takes place up to a 3 Si:1 Al ratio.

For a significant structural analysis of muscovite and phlogopite all the above factors must be accounted.

Ishii and co-workers (1967) in their normal coordinate treatment of the Si_2O_5 and $AlSi_3O_{10}$ layers, use the symmetry species and selection rules given in (Table XXXV) on the previous page.

By this treatment of the infrared data, they are able to assign or calculate the vibrational modes of the Si_2O_5 layer lattice (Fig. 7). Changes of the Si_2O_5 frequencies (C_{6v}) due to distortion of the Si_2O_5 layer (C_{3v}) and Al substitution may also be accounted for by this method (Fig. 8).

In order to investigate vibrations of the octahedral layers and potassium atoms between the silicate layers, the optically active lattice vibrations of idealized di- and trioctahedral type micas were calculated

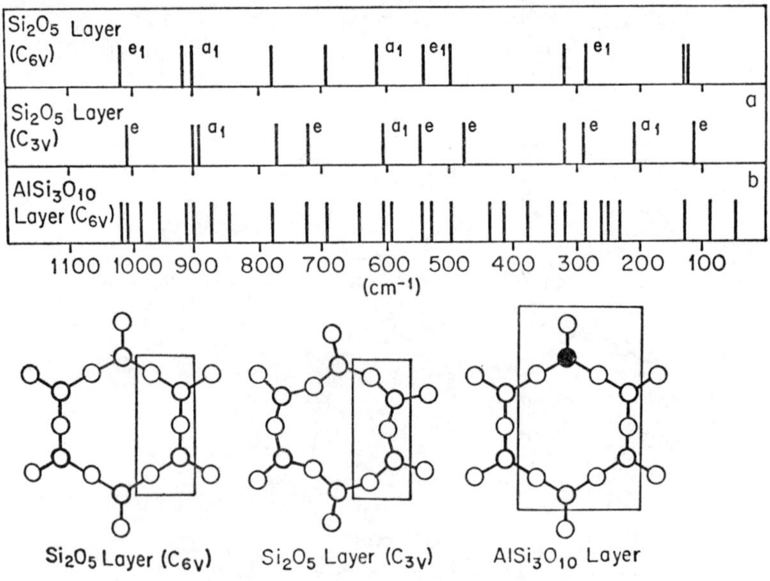

FIG. 8. Changes in Si_2O_5 frequencies due to, (a) distortion of the Si_2O_5 layer, and (b) due to aluminum substitution. O O, ○ Si, ● Al.

by normal coordinate treatment. In this calculation, it was assumed that the Si_2O_5 layer is not distorted and Al substitution does not take place; the stacking arrangement was assumed to be the M type. With modification of this calculation for Si_2O_5 ring deformation and Al substitution, assignment of some bands is permitted. Observed and calculated frequencies obtained by this method are illustrated in the paper by Ishii *et al.* (1967).

The interlayer vibrations were all found to lie below 250 cm^{-1}. Modes of interlayer vibrations (Fig. 9) and modes of the vibration of potassium atoms between the silicate layers of muscovite (Fig. 10) are given for the layer silicates.

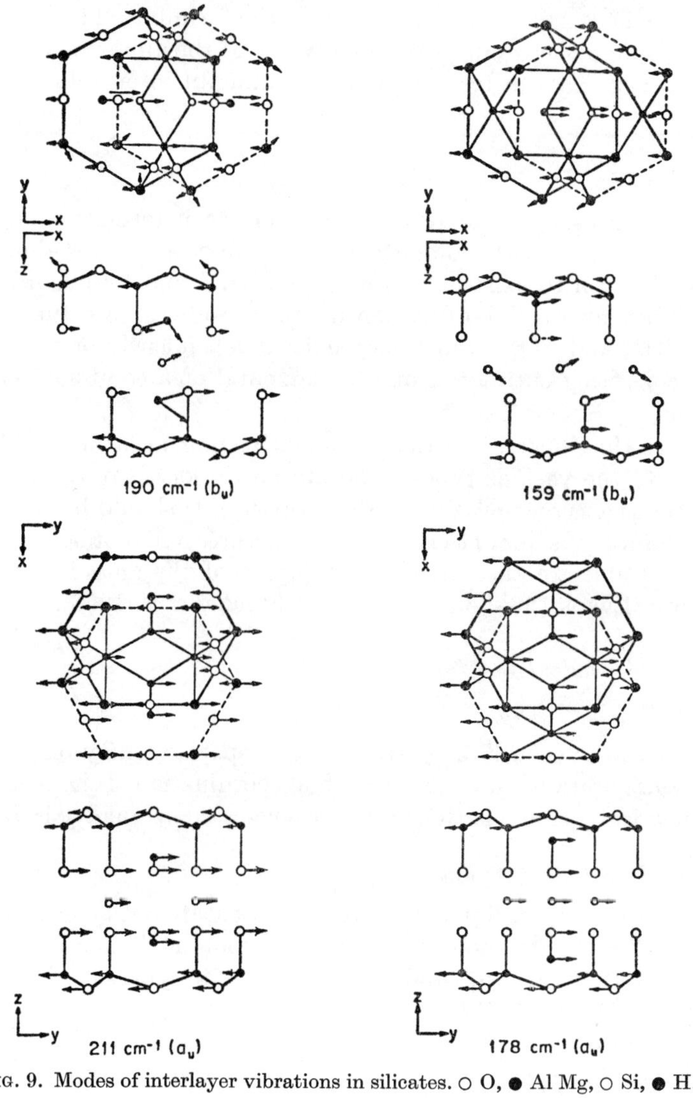

FIG. 9. Modes of interlayer vibrations in silicates. ○ O, ● Al Mg, ○ Si, ● H.

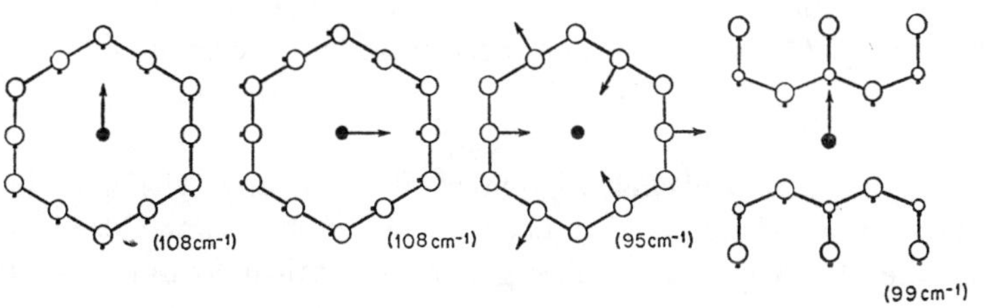

FIG. 10. Modes of vibration of the potassium atoms between the silicate layers of muscovite. ○ O, ○ Si, ● K.

For the most part, work on silicates in the far-infrared is very sparse, perhaps due to the complexities involved and the almost necessary use of normal coordinate treatment for meaningful interpretation of the spectra.

C. The Borates

The borates are similar to the silicates in their formation of anionic frameworks. Increased complexities arise in the borates due to the ability of boron to exist in both three-fold and four-fold coordination. This ability for dual coordination number results in two fundamental units, BO_3^{3-} and BO_4^{5-}, which may exist independently or form frameworks consisting entirely of one fundamental or a combination of the two units.

As with the silicates, a strictly descriptive nomenclature will be used to identify the various types of borate networks, thereby overcoming the variance of nomenclature in the literature. It should be emphasized at this point that some of the following structures have not been definitely confirmed and are theoretical assumptions of the anionic framework structure (Evans 1964; Moeller 1952; Schroeder *et al.* 1966).

1 *The Independents*

(a) *The Trigonal Anion*

The boron-oxygen linkage results in a separate entity of planar trigonal configuration due to the three-fold coordination of the boron atom. The chemical formula is BO_3^{3-} and the suggested symmetry is D_{3h}.

(b) *The Tetrahedral Anion*

The boron-oxygen linkage results in a separate entity of tetrahedral configuration due to the four-fold coordination of the boron atom. The chemical formula is BO_4^{5-} and the suggested site symmetry is T_d.

2. *Finite Linkage*

(a) *The Double Triangle*

Two adjacent trigonal groups share one oxygen atom. The chemical formula is $B_2O_5^{4-}$. It has low symmetry and has site group C_1.

(b) *The Ring Structure*

Two oxygen atoms of each trigonal group are shared with neighbouring trigonal groups, their angular position being such that closed units of a ring-like structure exist. The general chemical formula is $(BO_2^-)_n$.

Where $n = 3$, the suggested symmetry of the ring is D_{3h}; other n values are not common but can exist.

3. *Infinite Linkage*

The boron trigonal groups are joined by the sharing of two oxygen atoms to produce a chain of infinite length. The general chemical formula is $(BO_2)_n^{n-}$. Only the single chain structure is characterized with certainty. Double chain structures may exist in inorganic compounds containing the $B_4O_7^{2-}$ group (that is, borax $Na_2B_4O_7.10H_2O$); however, this has not been confirmed.

4. *Sheet Structures*

Basically, two-dimensional sheet structures form when borate rings contain a four coordinated boron.

(a) *The Double Ring*

Two, three-boron ($n = 3$) rings are coupled by means of a tetrahedral boron atom common to both rings. The general chemical formula of this anion is B_5O_8. Interconnection of these double rings is accomplished by a sharing of the external oxygen atoms to form extended two-dimensional $(x - y)$ sheets frequently finite in extension.

(b) *The Complex Ring*

The structure consists of two trigonal boron atoms and one tetrahedral boron atom joined by a sharing of oxygen atoms to form a six-membered ring (3B plus 3O). The ring is generally connected to similar units through the off-ring oxygen atoms resulting in a two-dimensional $(x - y)$ extension resembling a sheet-like structure.

5. *Three-dimensional Network*

(a) *The Tetrahedral Array*

Each boron tetrahedron group shares all of its oxygen atoms with other adjacent tetrahedra resulting in a complex three-dimensional array. The general chemical formula is B_2O_3 and the suggested site symmetry is C_{3v}.

(b) *The Ring Arrays*

(i) The eight-member ring. The anionic group consists of two BO_4 tetrahedra and two BO_3 triangles forming an eight-membered ring (4B plus 4O) in which the opposite tetrahedral BO_4 groups provide an oxygen bridge. The units are interconnected by oxygen atoms to form a three-dimensional network. The general chemical formula is $B_4O_7^{2-}$.

(ii) The six-member ring. The anion of this network contains rings of a structure similar to those described for finite structure. The anionic framework is three-dimensional resulting from two sets of rings connected by means of their off-ring oxygen atoms. The basic chemical formula is B_2O_3.

6. *Analogous Hydrated Forms*

It should be noted that, as in the silicates, the suggested symmetries may change depending on the chemical, polymorphic and isomorphic characteristics of each particular sample. The basic precept necessary for the interpretation of the infrared spectra of the six units is the same as those previously stated for the silicates.

Very little work has been done on the borates below 500 cm^{-1}. The work that has been done either leaves these bands unassigned or tentatively interprets the bands as M–O modes. Laperches and Tarte (1966) investigated low frequencies (below 500 cm^{-1}) in the infrared spectra of rare earth borates and have been able to establish some fundamental suggestions as to the interpretation of these low-lying modes. They explain the difference in frequency between $ScBO_3$ ($\sim$420 cm^{-1}) and $PrBO_3$ ($\sim$300 cm^{-1}) by the simultaneous intervention of two factors, a mass effect and the influence of the coordination number n (at. wt. for Sc = 45, Pr = 141; coordination number n for Sc = 6, Pr = 9), and logically concludes that the ScO_6 group frequencies should be higher. The mode suggested to produce the band is a librational movement of the BO_3 ion rendered possible by the force of repulsion created by the presence of the cation X^{3+}. This interpretation is identical to that proposed for the 320 cm^{-1} band in the calcite structure of calcium carbonate, movement of libration for the CO_7 group.

Apart from the intense absorption at $\sim$420 cm^{-1} in the spectrum of $ScBO_3$, weak bands at $\sim$263 cm^{-1} and 285 cm^{-1} are observed. No definite assignments are given but resonance frequencies are suggested. Laperches and Tarte also found when analysing some borates, the presence of isolated BO_4 groups considerably complicated the problem of band assignment in the region below 650 cm^{-1} due to overlap of absorptions due to isolated BO_4 groups and XO_8 groups. Other workers (Weir and Schroeder 1964) have attempted to distinguish these two groups by isotopic displacement of the boron (B^{10}–B^{11}) which causes shifts in the boron frequencies and leaves the XO_8 frequencies undisturbed. The most use of this method has been made in the mid-infrared region.

Laperches and Tarte (1966) were able to determine that frequencies due to XO_8 vibrations associated with borates are a function of cation size of rare earth element above 650 cm^{-1} (essentially harmonic oscillation of X–O bonds). They reasoned this to be due to the existence of two different bond lengths in the XO_8 coordination group (2.21A and 2.41A for YBO_3).

The infrared spectra of the hydrated borates (Weir 1966) are more complex than those of the anhydrous types due to the effects of the hydration of the borate anions and hydrogen bonding between anions. It can be concluded that the infrared spectra of the hydrated borates are not of acceptable reliability at the present time. The lack of reproducibility is most probably to be attributed to differences in chemical composition and degree of crystallinity and to varying interactions between the hydrated borates and the mulling and/or sampling technique. Any attempt to understand the spectra of the hydrated borates will probably require deuteration as well as boron isotope substitution.

A listing of low-lying frequencies found in borates investigated by various workers is found in the Appendix.

An investigation of borates similar to the one accomplished by Ishii *et al.* (1967) on the silicates would be valuable. Absorptions due to lattice vibrations of the anionic framework, cationic-anionic interaction, lattice distortions, isomorphism and polymorphism presumably should be evidenced for the borates in the far-infrared region. In fact, borate spectra containing large numbers of bands, particularly below 600 cm^{-1}, may be taken to be indicative of structurally complex anions of low symmetry. The same principles of investigation, structural or crystallographic knowledge and normal coordinate treatment used for the silicates would undoubtedly be of benefit in the study of the borates.

VII. Reviews

A number of reviews and articles dealing specifically with inorganic applications of far-infrared spectroscopy have now appeared; some have already been mentioned elsewhere in the text.

A very useful general introduction to vibrational spectra, especially as applicable to inorganic systems, has been given by Ware (1968). The well-known text of Nakamoto (1963) gives a theoretical background together with extensive data on vibrational spectra of inorganic and coordination compounds. More recently a very comprehensive compilation of data on the subject of metal-ligand vibrations together with discussion of various trends and correlations has been given by Adams (1967). Many useful data and applications of far-infrared spectroscopy, in particular inorganic applications, are contained in a review by Brasch *et al.* (1968). Reviews by Clark (1965a and 1967) are mainly concerned with far-infrared spectra of transition metal systems. Much useful data on the vibrational spectra of transition metal complexes are summarized in another review, together with discussion on the nature of metal-ligand bonds (James and Nolan 1968). Low-frequency vibrational spectra of some inorganic systems are included in a report by Hendra (1968). A discussion of the significance and application of a general quadratic valence force field applied to square planar complexes has been given in a review by Jones (1966).

References

Adams, D. M. (1967). "Metal-Ligand and Related Vibrations." Edward Arnold, London.

Adams, D. M. and Chandler, P. J. (1965). *Chemy Ind.* 269.

Adams, D. M. and Chandler, P. J. (1966). *Chem. Commun.* 69.

Adams, D. M. and Chandler, P. J. (1967). *J. chem. Soc.* (A) 1009.

Adams, D. M. and Churchill, R. G. (1968). *J. chem. Soc.* (A) 2141.

Adams, D. M. and Cornell, J. B. (1967). *J. chem. Soc.* (A) 884.

Adams, D. M. and Gebbie, H. A. (1963). *Spectrochim. Acta* **19**, 925.

Adams, D. M. and Lock, P. J. (1967). *J. chem. Soc.* (A) 620.

Adams, D. M. and Morris, D. M. (1967a). *J. chem. Soc.* (A) 1666.

Adams, D. M. and Morris, D. M. (1967b). *J. chem. Soc.* (A) 1669.

Adams, D. M. and Morris, D. M. (1968). *J. chem. Soc.* (A) 694.

Adams, D. M. and Newton, D. C. (1968). *J. chem. Soc.* (A) 2262.

Adams, D. M., Gebbie, H. A., and Peacock, R. D. (1963a). *Nature, Lond.* **199**, 278.

Adams, D. M., Chatt, J., Davidson, J. M., and Gerratt, J. (1963b). *J. chem. Soc.* 2189.

Adams, D. M., Chatt, J., Gerratt, J., and Westland, A. D. (1964). *J. chem. Soc.* 734.

Adams, D. M., Chandler, P. J., and Churchill, R. G. (1967). *J. chem. Soc.* (A) 1272.

Allen, G. and Warhurst, E. (1958). *Trans. Faraday Soc.* **54**, 1786.

Allkins, J. R. and Hendra, P. J. (1966). *Spectrochim. Acta* **22**, 2075.

Allkins, J. R. and Hendra, P. J. (1967a). *Spectrochim. Acta* **23A**, 1671.

Allkins, J. R. and Hendra, P. J. (1967b). *J. chem. Soc.* (A) 1325.

Allkins, J. R. and Hendra, P. J. (1968). *Spectrochim. Acta* **24A**, 1305.

Avery, J. S., Burbridge, C. D., and Goodgame, D. M. L. (1968). *Spectrochim. Acta* **24A**, 1721.

Bagnall, K. W. and Brown, D. (1964). *J. chem. Soc.* 3021.

Bagnall, K. W., Brown, D., and du Preez, J. G. H. (1964). *J. chem. Soc.* 2603.

Balls, A., Downs, A. J., Greenwood, N. N., and Straughan, B. P. (1966). *Trans. Faraday Soc.* **62**, 521.

Beattie, I. R. and Rule, L. (1964). *J. chem. Soc.* 3267.

Beattie, I. R., Gilson, T. R., and Cocking, P. (1967a). *J. chem. Soc.* (A) 702.

Beattie, I. R., Gilson, T. R., Livingston, K., Fawcett, V., and Ozin, G. (1967b). *J. chem. Soc.* (A) 712.

Beattie, I. R., McQuillan, G. P., Rule, L., and Webster, M. (1963). *J. chem. Soc.* 1514.

Bennett, M. A. and Clark, R. J. H. (1964). *J. chem. Soc.* 5560.

Bennett, M. A., Clark, R. J. H., and Milner, D. L. (1967). *Inorg. Chem.* **6**, 1647.
Berry, L. G. and Mason, B. (1959). "Mineralogy." Freeman, San Francisco, California.
Boorman, P. M. and Straughan, B. P. (1966). *J. chem. Soc.* (A) 1514.
Bradford, C. W., van Bronswyk, W., Clark, R. J. H., and Nyholm, R. S. (1968). *J. chem. Soc.* 2456.
Brasch, J. W., Mikawa, Y., and Jakobsen, R. J. (1968). "Applied Spectroscopy Reviews," Vol. 1. (Brame, E. G. ed.) Edward Arnold, London.
Brier, P. N., Chalmers, A. A., Lewis, J., and Wild, S. B. (1967). *J. chem. Soc.* (A) 1889.
Brown, M. P., Cartmell, E., and Fowles, G. W. A. (1960). *J. chem. Soc.* 506.
Carey, N. A. D. and Clark, H. C. (1967). *Chem. Commun.* 292.
Carey, N. A. D. and Clark, H. C. (1968). *Inorg. Chem.* **7**, 94.
Chantry, G. W., Finch, A., Gates, P. N., and Steele, D. (1966). *J. chem. Soc.* 896.
Clark, R. J. H. (1965a). *Rec. chem. Progr.* **26**, 269.
Clark, R. J. H. (1965b). *Spectrochim. Acta* **21**, 955.
Clark, R. J. H. (1967). *In* "Halogen Chemistry." (Gutman, V. ed.) Vol. **3**, p. 85. Academic Press, London.
Clark, R. J. H. and Dunn, T. M. (1963). *J. chem. Soc.* 1198.
Clark, R. J. H. and Williams, C. S. (1965). *Inorg. Chem.* **4**, 350.
Clark, R. J. H., Kepert, D. L., Nyholm, R. S., and Rodley, G. A. (1966). *Spectrochim. Acta* **22**, 1697.
Clark, R. J. H., Davies, A. G., and Puddephatt, R. J. (1968a). *J. chem. Soc.* 1828.
Clark, R. J. H., Maresca, L., and Puddephatt, R. J. (1968b). *Inorg. Chem.* **7**, 1603.
Coates, G. E. and Parkin, C. (1963). *J. chem. Soc.* 421.
Coates, G. E. and Ridley, D. (1964). *J. chem. Soc.* 166.
Corbett, J. D. (1962). *Inorg. Chem.* **1**, 700.
Corey, E. R. and Dahl, L. F. (1962). *Inorg. Chem.* **1**, 521.
Cotton, F. A. (1966). *Q. Rev. chem. Soc.* **20**, 389.
Cotton, F. A. and Wing, R. M. (1965). *Inorg. Chem.* **4**, 1328.
Cotton, F. A., Wing, R. M., and Zimmerman, R. A. (1967). *Inorg. Chem.* **6**, 11.
Davies, J. E. D. and Long, D. A. (1968). *J. chem. Soc.* (A) 1761.
Deacon, G. B. and Green, J. H. S. (1968a). *Spectrochim. Acta* **24A**, 845.
Deacon, G. B. and Green, J. H. S. (1968b). *Spectrochim. Acta* **24A**, 959.
Decius, J. C. (1965). *Spectrochim. Acta* **21**, 15.
Drago, R. S. (1965). "Physical Methods in Inorganic Chemistry." Reinhold Publishing Corp., New York.
Edwards, H. G. M., Ware, M. J., and Woodward, L. A. (1968). *Chem. Commun.* 540.
Evans, R. C. (1964). "Crystal Chemistry" 2nd Edition. Cambridge University Press, Cambridge.
Fahrenfort, J. (1963). *In* "Infrared Spectroscopy and Molecular Structure." Davies, Mansel ed.) p. 377. Elsevier, Amsterdam.
Fateley, W. G., Matsubara, I., and Witkowski, R. E. (1964). *Spectrochim. Acta* **20**, 1461.
Frankiss, S. G., Miller, F. A., Stammreich, H., and Sans, T. T. (1967). *Spectrochim. Acta* **23A**, 543.
Fujita, J. and Nakamoto, K. (1964). *Bull. chem. Soc. Japan* **37**, 528.
Gager, H. M., Lewis, J., and Ware, M. J. (1966). *Chem. Commun.* 616.
Goodfellow, R. J., Goggin, P. L., and Venanzi, L. M. (1967). *J. chem. Soc.* (A) 1897.

Green, J. H. S., Kynaston, W., and Rodley, G. A. (1968). *Spectrochim. Acta* **24A**, 853.

Greenwood, N. N., Prince, D. J., and Straughan, B. P. (1968). *J. chem. Soc.* 1694.

Hartley, D. and Ware, M. J. (1967). *Chem. Commun.* 912.

Hartley, D., Kilty, P. A., and Ware, M. J. (1968). *Chem. Commun.* 49.

Hendra, P. J. (1968). *In* "Annual Reports on the Progress of Chemistry for 1967", Vol. 64A. The Chemical Society, London.

Hendra, P. J. and Jovic, Z. (1968). *Spectrochim. Acta* **24A**, 1713.

Herzberg, G. (1950). "Spectra of Diatomic Molecules." Van Nostrand, New York.

Horner, S. M., Clark, R. J. H., Crociani, B., Copley, D. B., Horner, W. W., Collier, F. N., and Tyree, S. Y. (1968). *Inorg. Chem.* **7**, 1859.

Hunt, G. R., Perry, C. H., and Ferguson, J. (1964). *Phys. Rev.* **134A**, 688.

Ishii, M., Shimanouchi, T., and Nakahire, M. (1967). *Inorg. Chim. Acta* **1**, 387.

James, D. W. and Nolan, M. J. (1968). *In* "Progress in Inorganic Chemistry." (Cotton, F. A. ed.) Vol. 9, p. 195. Interscience, New York.

Janz, G. J. and James, D. W. (1963a). *J. Chem. Phys.* **38**, 902; (1963b). *J. Chem. Phys.* **38**, 905.

Jones, L. H. (1962a). *J. Chem. Phys.* **36**, 1209.

Jones, L. H. (1962b). *J. molec. Spectrosc.* **8**, 105.

Jones, L. H. (1966). *Co-ord. Chem. Rev.* **1**, 351.

Laperches, J. P. and Tarte, P. (1966). *Spectrochim. Acta* **22**, 1201.

Lewis, J., Manning, A. R., Miller, J. R., Ware, M. J., and Nyman, F. (1965a). *Nature, Lond.* **207**, 142.

Lewis, J., Nyholm, R. S., and Rodley, G. A. (1965b). *J. chem. Soc.* 1483.

Long, D. A. and Chau, J. Y. H. (1962). *Trans. Faraday Soc.* **58**, 2325.

Mackay, R. A. and Schneider, R. F. (1968). *Inorg. Chem.* **7**, 454.

Mills, I. M. (1963). *In* "Infrared Spectroscopy and Molecular Structure." (Davies, Mansel ed.) p. 166. Elsevier, Amsterdam.

Moeller, T. (1952). "Inorganic Chemistry." John Wiley, New York.

Nakamoto, K. (1963). "Infrared Spectra of Inorganic and Coordination Compounds." John Wiley, New York.

Nakamoto, K., Fujita, J., Condrate, R. A., and Morimoto, Y. (1963). *J. Chem. Phys.* **39**, 423.

Oertel, R. P. and Plane, R. A. (1967). *Inorg. Chem.* **6**, 1960.

Onishi, T. and Shimanouchi, T. (1964). *Spectrochim. Acta* **20**, 325.

Poulet, H. and Mathieu, J. P. (1963). *J. Chim. phys.* **60**, 442.

Quicksall, C. O. and Spiro, T. G. (1967). *Chem. Commun.* 839.

Sabatini, A. and Sacconi, L. (1964). *J. Am. Chem. Soc.* **86**, 17.

Sabatini, A., Sacconi, L., and Schettino, V. (1964). *Inorg. Chem.* **3**, 1775.

Schroeder, R. A., Lippincott, E. R., and Weir, C. E. (1966). *J. inorg. nucl. Chem.* **28**, 1397.

Shimanouchi, T. (1963). *Pure appl. Chem.* **7**, 131.

Shimanouchi, T. and Hiraishi, J. (1966). *Spectrochim. Acta* **22**, 1483.

Shimanouchi, T. and Nakagawa, I. (1962). *Spectrochim. Acta* **18**, 89.

Shimanouchi, T. and Nakagawa, I. (1964). *Inorg. Chem.* **3**, 1805.

Shimanouchi, T. and Nakagawa, I. (1966a). *Spectrochim. Acta* **22**, 759.

Shimanouchi, T. and Nakagawa, I. (1966b). *Spectrochim. Acta* **22**, 1707.

Shimanouchi, T., Hiraishi, J., and Nakagawa, I. (1964). *Spectrochim. Acta* **20**, 819.

Shimanouchi, T., Nakagawa, I., Hiraishi, J. and Ishii, M. (1966). *J. molec. Spectrosc.* **19**, 78.

Shimanouchi, T., Hiraishi, J., and Nakagawa, I. (1968). *Spectrochim. Acta* **24A**, 819.

Stammreich, H. and Sans, T. T. (1967). *J. molec. Struct.* **1**, 55.

Stammreich, H., Kawai, K., Sala, O., and Krumholz, P. (1961). *J. Chem. Phys.* **35**, 2175.

Taimsalu, P. and Wood, J. L. (1964). *Spectrochim. Acta* **20**, 1357.

Ware, M. J. (1968). *In* "Physical Methods in Advanced Inorganic Chemistry." (Hill, H. A. O. and Day, P. eds.) p. 231. Interscience, New York.

Watters, K. L., Brittain, J. N., and Risen, W. M. (1969). *Inorg. Chem.* **8**, 1347.

Wei, C. H. and Dahl, L. F. (1966). *J. Am. chem. Soc.* **88**, 1821.

Weidlein, J. and Dehnicke, K. (1965). *Z. anorg. allg. Chem.* **337**, 113.

Weir, C. E. (1966). *J. Res. natn Bur. Stand.* **70A**, 153.

Weir, C. E. and Schroeder, R. A. (1964). *J. Res. natn Bur. Stand.* **68A**, 465.

Wendlandt, W. W. and Hecht, H. G. (1966). "Reflectance Spectroscopy." Interscience, New York.

Wharf, I. and Shriver, D. F. (1969). *Inorg. Chem.* **8**, 914.

Wilmshurst, J. K. (1960). *J. molec. Spectrosc.* **5**, 343.

Wilson, E. B., Decius, J. C., and Cross, P. C. (1955). "Molecular Vibrations." McGraw-Hill, New York.

Woodward, L. A. (1934). *Phil. Mag.* **18**, 823.

Woodward, L. A. (1958). *Trans. Faraday Soc.* **54**, 1271.

Woodward, L. A. and Taylor, M. J. (1960). *J. chem. Soc.* 4473.

Chapter 7

METAL-ORGANIC AND ORGANIC COMPOUNDS

I. Introduction

The term metal-organic is differently understood by different authors, and is here intended to represent compounds in which a carbon atom of some moiety, which would generally be conceded by chemists to be organic, is directly linked by a largely covalent bond to a metal or metalloid. Coordination and addition compounds of various types, in which the organic grouping is a ligand, are excluded for purely practical reasons, and discussion of some of these are given in Chapter 6.

Metal-carbon stretching modes in metal-organic compounds will not generally be expected, on mass considerations, to lie in the far infrared. Considering the very heavy metal lead (atomic mass 207.2), it is found that the carbon-lead stretching frequency lies between 420 and 480 cm^{-1}. Consequently, unless accompanied by very low force constants, very few metal-carbon stretches will be observed in the low-frequency range. On the other hand, bending and deformation modes generally will often lie in this region. For these and for organic compounds in general, far infrared spectroscopy will have value in elucidating structure (including isomeric problems), evaluating force constants, and confirming or establishing fundamental frequencies of vibration from which thermodynamic parameters may be computed. It is impossible to attempt a complete coverage, but some examples illustrating such applications follow.

One of the greatest practical uses of conventional mid- and near-infrared spectroscopy is the recognition of specific chemical groupings, for example C–H; C≡N; O–H, and so on, by the position, and sometimes intensity or shape, of their absorptions in the complex spectrum of the compound in which they occur. Such recognizable absorptions are termed characteristic or group frequencies, and it is assumed that the modes of which they are a manifestation are relatively "pure", and are hence transferable from one compound to another. The establishment and testing of such characteristic frequencies, principally for organic compounds, has attracted enormous attention over many years, and is extensively documented (see, for example, Bellamy 1962, 1968). A comprehensive collection of spectra of a very large number of compounds, and consideration of their possible use in establishing characteristic frequencies and correlations to 300 cm^{-1}, has recently been published by Bentley *et al.* (1968).

The factors which are of importance in determining whether or not a given vibration is likely to be characteristic have been discussed by many authors, (for example, Bellamy *vide supra*; Lord and Miller 1956) and are summarized by Bentley. Two conditions are that the absorption must clearly be of sufficient intensity to permit easy identification in a spectrum, and also that the vibration should preferably not mix to any great extent with other bond vibrations of appropriate symmetry. Both requirements become less easily satisfied in the low-frequency region. This does not mean that qualitative analysis using far-infrared spectroscopy is impossible. However, considerable experience and judgment may be needed in making evaluations, and fruitful results are most likely to be obtained if the spectra in higher-frequency ranges are used in conjunction with those covering the low-frequency region. Successful examples of this are given by Bentley, using spectra obtained in the caesium bromide region, for example, to analyse paints.

A clear discussion which emphasizes some of the above points has been given by Stewart (1965) in a consideration of the use of far-infrared bending modes of various organic molecules as characteristic group frequencies. Citing spectra obtained by Hidalgo (1962, 1958) for aliphatic and aromatic nitriles, it is shown that the nitrile bending mode is apparently located at 371 ± 11 cm^{-1} in aliphatic molecules, and thus does appear as a characteristic frequency. With aromatic nitriles, however, the limit is much greater, and is put at 362 ± 51 cm^{-1}, while in the trihalomethyl nitriles (X_3CCN) absorptions at 211 ± 48 cm^{-1} are found. In the first place, it is uncertain whether observed absorptions can reasonably be ascribed solely to particular modes such as bending in these molecules. Secondly, assignments in some aromatic compounds are often especially difficult; this may be illustrated by noting that it is possible for two papers dealing with aromatic compounds, published in the same issue of *Spectrochimica Acta* in 1968, to disagree both with each other and with a third paper in an issue a few months later. Bending frequencies associated with acetylenic groups, the –C≡CH bend and the –C–C≡C– bend, have been discussed by Nyquist and Potts (1960) and are again summarized by Stewart. Some results are given in Table I.

TABLE I. Bending Assignments in Methylacetylenes[a] (cm^{-1}).

Mode[b]	CH_3CCH	CH_3CCCl	CD_3CCCl	CH_3CCBr	CH_3CCI
C–C–C bend (ν_4)	336	358	343	343	343
C–C–X bend (ν_5)	642	184	(173)	171	163

[a] Values from Daykin *et al.* (1962). [b] The term "mode" is used as an abbreviated form of "approximate description of mode".

The ν_4 –C≡C–C bend is remarkably constant along the series considered, and is in agreement with normal coordinate calculations (Daykin *et al.* 1962). The –C≡C– halogen bend, ν_5, appears to be largely independent of the nature of the halogen and might be diagnostic. However, the introduction of additional –C≡C– moieties into a molecule in the presence of a –C–N group causes complications. Bentley *et al.* (1968) include an extensive compilation of data for both nitriles and alkynes.

II. Lead Compounds

Studies on organo-lead compounds provide an excellent example of the use of far-infrared spectroscopy. Previous low-frequency assignments for the simple aliphatic molecules, for example tetramethyl lead, relied on Raman measurements. These are complicated by the well-known susceptibility of such compounds to photolytic decomposition, a difficulty which tends to be exacerbated, rather than removed, by the use of modern laser sources. Further, Raman assignments are almost invariably on condensed phases, and the use of Raman-measured values in tetramethyl lead (*vide infra*) has led to a discrepancy in the value of the entropy.

A. Tetramethyl Lead

This has been the subject of numerous mid-infrared and Raman studies, for example by Lippincott and Tobin (1953), Sheline and Pitzer (1950), and several other pertinent references are quoted by Scott *et al.* (1965); complete vibrational assignment had been given without measurement in the far-infrared region. However, such assignments were inconsistent with the precisely known value of the entropy in the gas phase, $S^\circ_{928.15} = 100.48 \pm 0.20$ cal deg^{-1} mole^{-1} (Scott *et al.* 1959); assuming free rotation of the methyl groups, the value calculated from Lippincott and Tobin's assignment is $S^\circ_{298.15} = 99.86$ cal deg^{-1} mole^{-1}. Available evidence (Jones and Sheppard 1965) suggests that any rotation barrier is likely to be very small ($<RT$), but in any event allowance for such effects would *increase* the discrepancy.

Two modes are found in the low-frequency region, the two skeletal C–Pb–C bends, ν_8 (t_2) and ν_{19} (e), and they were previously assigned to a single, very broad, asymmetric band with maximum intensity at about 132 cm^{-1} in the Raman spectrum. With T_d symmetry both modes are Raman-active, but only the t_2 species is permitted by the selection rules in the infrared. Previous entropy discrepancies for toluene (Scott *et al.* 1962) had been resolved by establishing that some liquid-state frequency values were significantly less than those found from gas-phase measurements. A far-infrared study of lead tetramethyl was therefore made by Scott *et al.* (1965, *vide supra*), using both liquid and gas samples. The spectra obtained showed only a single, sharp absorption in the liquid state at 130 cm^{-1} (as with the Raman value), and in the gas phase at the considerably lower position of 120 cm^{-1}.

Unfortunately, it was not possible to locate in the infrared the position of the other low-frequency bend, ν_{19}, through combination bands. However, knowing the corrected t_2 fundamental, the usual calculation was reversed and the *e* mode calculated using the observed accurate entropy values and conventional statistical procedures. Anharmonicity was neglected, no rotational contribution from the methyl group was assumed, and the value obtained for the *e* mode was 117.6 ± 6 cm^{-1}. A normal coordinate analysis was given, and the complete thermodynamic functions computed from 0–1000°K.

B. Tetraethyl Lead

This molecule has been studied in the far infrared by Wood and Taimsalu (1963) to 55 cm^{-1}. As noted for tetraethyl tin, replacement of methyl by ethyl gives rise to three skeletal frequencies not observed in the corresponding tetramethyl compound. One of these, the Pb–C–C bending mode, was assigned to a strong absorption at 240 cm^{-1}. The torsion was not observed; the barrier to rotation is probably very low and is expected to lie below the limit of measurement (55 cm^{-1}). Assignments of the C–Pb–C bending modes, together with other low-frequency data for tetramethyl lead and germanium are shown in Table II.

TABLE II. Low-Frequency Assignments in Tetramethyl Lead, Tetramethyl Germanium, and Tetraethyl Lead (cm^{-1}).

Mode	$Pb(CH_3)_4$	$Ge(CH_3)_4$	$Pb(C_2H_5)_4$
X–C–C bend			240 s (liq.)
C–X–C bend (ν_8)	120 (gas) 130 (liq.)	175 R (liq.)	86 m; 96.5 R (liq.)
C–X–C bend (ν_{19})	118 (gas, liq.)	195 R (liq.)	122 m (liq.)

Raman results are from Lippincott and Tobin.

C. Mixed Alkyl Compounds

Jackson and Nielsen (1964) have developed a technique by which photo-sensitive compounds may be examined under Raman conditions, by working at low temperatures. This has been exploited for use with tetramethyl lead, methyltriethyl lead, trimethylethyl lead, dimethyldiethyl lead, tetraethyl lead, tetraethyl lead-d_{20} and also triethyl bismuth. It has hence been possible to revise and extend assignments for some of these and to give partial interpretations of the higher alkyls.

III. Aluminium Compounds

A. Trimethylaluminium

This compound, the simplest aliphatic organo-aluminium compound, has been extensively investigated in all regions of the infrared, and

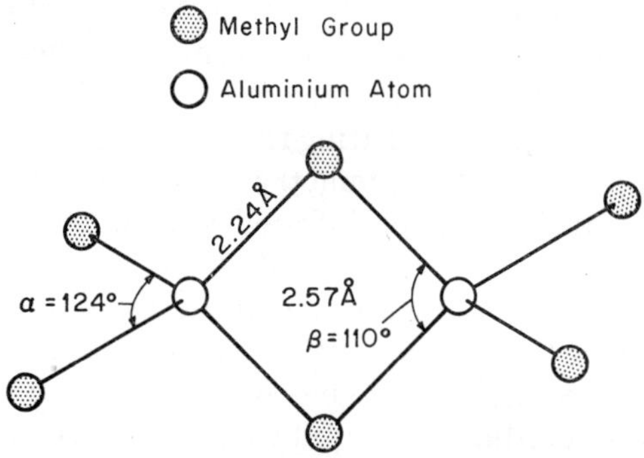

FIG. 1. Structure of trimethylaluminium.

also the Raman, for several years. The interest depends at least in part on its structure, accurately known from X-ray analysis by Rundle and Lewis (1953), which resembles that of diborane, and not trimethylboron, and is shown in Fig. 1. The bridged structure is that of a typical electron-deficient molecule, and the aluminium–aluminium distance, 2.57 Å, is close to the sum of the Pauling tetravalent covalent radii, 2.52 Å, so that the possibility of a metal–metal bond is likely.

TABLE III. Assignments for Trimethylaluminium (cm^{-1}).

	Species Mode	Observed	Calculated
Raman	a_g ν_1 stretch	592	633
	ν_2 stretch	453	473
	ν_3 ring	313	305
	ν_4 bend	153	145
	b_{1g} ν_6 stretch	—	395
	ν_7 wag	—	210
	b_{2g} ν_{11} stretch	683	690
	ν_{12} rock	—	173
	b_{3g} ν_{15} twist	185?	186
I.R. active	b_{1u} ν_8 stretch	697	688
	ν_9 rock	—	257
	ν_{10} puckering	—	48
	b_{2u} ν_{13} stretch	480	478
	ν_{14} wag	175	173
	b_{3u} ν_{16} stretch	616	585
	ν_{17} stretch	367	358
	ν_{18} bend	175	183

Ogawa *et al.* (1965) have measured the infrared spectrum in the liquid state to 65 cm^{-1} and have also given a normal coordinate analysis using a procedure similar to that adopted previously for diborane (Ogawa and Miyazawa 1964). Approximating the methyl groups as point masses, the molecule possesses D_{2h} symmetry, three of the eight symmetry species (b_{1u}, b_{2u}, and b_{3u}) being infrared active. Two absorptions only were found in the far infrared, at 367 and 175 cm^{-1}, the only mode expected below 65 cm^{-1} being the ring puckering, ν_{10}. In addition to the results obtained in this study, several other measurements, including spectra of the deuterated compound by Gray (1963), were considered with the normal coordinate computations; the recommended assignments are as shown in Table III. The stretching force constant for the Al–Al bond was calculated to be 1.24 md $Å^{-1}$, consistent with some bond interaction.

B. Methyl Aluminium Chlorides

Gray (1963) has studied to 300 cm^{-1} the infrared spectra of methylaluminium dichloride, dimethyl aluminium chloride and deuterated dimethyl aluminium chloride. Trimethyl aluminium chloride and its deuterated analogue were also measured, and their assignments were considered by Ogawa (*vide supra*). Dimethyl aluminium chloride is known by electron diffraction measurements (Brockway and Davidson 1941) to be dimeric in the gas phase and it is established that the bridging is through the chlorine atoms, and not the methyl groups as in the trimethyl dimer. Raman and infrared data for this and analogous molecules are available from Hoffmann (1960), Groenewege (1958, 1959) and van der Kelen and Herman (1956), but spectra were not all in agreement. Gray was able to resolve the discrepancies between reported spectra by the suggestion that some samples, especially that of van der Kelen and Herman, were contaminated with $(CH_3)_3Al_2Cl_3$, the so-called sesquichloride, which is the initial product in the preparation of dimethyl aluminium chloride from methyl chloride and aluminium. Only two absorptions lower than 400 cm^{-1} were found, a strong band at 310 cm^{-1} and a medium intensity band at 343 cm^{-1}.

C. Methylaluminium Dichloride

Like the corresponding compounds above, this is dimeric and bridged through the chlorine atoms. *Cis* and *trans* isomers are possible: the *trans*-1,2-dichloro and *cis*-1,2-dichloro forms, of respective C_{2h} and C_{2v} symmetries. Consideration of the selection rules and available spectra, while not complete, apparently suggest the *trans* configuration. Recommended assignments are shown in Table IV.

TABLE IV. Low-Frequency Assignments in Methylaluminium Chlorides (cm^{-1}).

Mode	$Al_2Cl_2(CH_3)_4$	$Al_2Cl_2(CD_3)_4$	$Al_2Cl_4(CH_3)_2$
Bridge stretch, ν_{17}	310	307	310
$AlCl_2$ bend, ν_{18}	343	331	

IV. Titanium Compounds

There are few reported vibrational studies of transition metal alkyls in general, doubtless owing to the notorious instability of these compounds.

A. Methyl Titanium Trichloride

Infrared studies on this compound have been reported by Karapinka *et al.* (1961) and Groenwege (1959) in the liquid and solution phases. Gray has made an infrared study in the vapour phase to 300 cm^{-1} and has also attempted to record spectra of CD_3TiCl_3. It was shown that previously reported absorption at 490 cm^{-1} assigned to the Ti–Cl antisymmetrical stretch in $(CH_3)TiCl_3$ was due to $TiCl_4$ impurity, and no absorption in this position was found in the vapour spectrum. It is assumed that the molecule is tetrahedral and of C_{3v} symmetry and the symmetrical Ti–Cl_3 stretch is assigned to a band at 391 cm^{-1} in the Raman spectrum. A weak band at 355 cm^{-1} was tentatively identified in the spectrum of a mixture containing CD_3TiCl_3, although the compound was clearly unstable (Table V).

TABLE V. Infrared Absorptions in Methyl Titanium Chloride (cm^{-1}).

$TiCl_3CH_3$	$TiCl_2(CH_3)_2$	$TiCl_3CD_3$
1380 w	1380 w	
550 s	550 s	530 s
464 s	462 m	405 m
391 w	427 w	
	396 w	

B. Dimethyl Titanium Dichloride

The first infrared spectrum of this was also reported by Gray (*vide supra*) who states that the spectrum is qualitatively in agreement with the expected tetrahedral structure.

V. Selenium Compounds

As might be expected, partly from their extremely poisonous and unpleasant nature, there has been little spectroscopic effort with compounds of this kind.

A. Methylselenol

A detailed vibrational study of this compound, and of CD_3SeH, CD_3SeD and CH_3SeD_3, both Raman and infrared, has been reported by Harvey and Wilson (1966) who give a complete assignment and a normal coordinate analysis. Assuming C_s symmetry by extrapolation from the structures of methanol and also methanethiol, all twelve normal modes should be active in both the Raman and the infrared. Low-frequency interests centre on the torsion of the SeH moiety against the methyl group, which should appear in the far infrared. In methylselenol a weak, featureless band was observed at about 145 ± 40 cm^{-1}. The very large limit of error is possibly due to overlapping by pure rotational bands in this region of the spectrum, and hence definition of the $0 \rightarrow 1$ torsional transition is very difficult. Using the procedure of Fateley and Miller (1963), and assuming a three-fold cosine barrier, leads to a value for the torsional barrier of 320 ± 40 cm^{-1}, which compares with the microwave value (Wilson and Thomas, Harvey and Wilson, *vide supra*) of about 350 cm^{-1}. A note on the relative magnitudes of this barrier compared to those of methyl mercaptan (444 cm^{-1}) and methanol (375 cm^{-1}) is made in Chapter 4.

VI. Mercury Compounds

A very recent comprehensive paper by Green (1968), in addition to presenting new infrared vibrational data over a very large range of the spectrum, gives a thorough discussion of much previous work.

A. Dimethyl Mercury

The simplest compound $Hg(CH_3)_2$ has, with a linear C–Hg–C skeleton, an effective point group D_{3d} (Bauman 1956) with no predicted infrared and Raman coincidences. Woodward (1963) has, however, confirmed the skeletal bending fundamental (class e_u) both in the Raman and the infrared near 160 cm^{-1} by direct measurement. The accepted vibrational assignment (Gutowsky 1949) is as shown in Table VI.

TABLE VI. Low-Frequency Modes and Force Constants in Dimethyl Mercury (cm^{-1}).

Species	Mode		Force constant (md $Å^{-1}$)
ν_3	C–Hg–C stretch	515	2.3
ν_{11}	C–Hg–C bend	156	0.09

B. Methyl Mercuric Cyanide

Hall and Mills (1966) have investigated the infrared spectrum of this compound to 300 cm^{-1}, and also reported the Raman spectrum. Assuming again a linear C–Hg–C skeleton, a normal coordinate calculation was made, from which the following assignments and modal descriptions of the low-frequency range were proposed.

TABLE VII. Low-Frequency Assignments and Force Constants in $(CH_3)HgCN$ (cm^{-1}).

Species	Mode	ν	Force constant (md $Å^{-1}$)
$\nu_5(a_1)$	Hg–CN stretch	386	2.1
$\nu_9(e)$	Hg–C–N bend	306	0.3
$\nu_{10}(e)$	C–Hg–C bend	75	0.15

It will be noted that the force constant values do not differ appreciably from those given for dimethyl mercury above.

C. Methyl Mercuric Halides

Green (1968) has reported infrared spectra to 40 cm^{-1} for $(CH_3)HgX$ (X = Cl, Br, I) and completed a normal coordinate calculation for the iodide, assuming C_{3v} symmetry. Observed frequencies, shown in Table VIII, are in good agreement with those calculated, and the description of modes in terms of conventional group and skeletal vibrations is satisfactory.

TABLE VIII. Low-frequency Modes for Methyl Mercuric Halides (cm^{-1}).

Species	$IHgCH_3$	$BrHg(CH_3)$	$ClHg(CH_3)$	Description
$\nu_4(a_1)$	184, 169	214	289	Hg–X stretch
$\nu_8(e)$	64	—	—	C–Hg–I def.

D. Ethyl Mercuric Bromide

Infrared measurements for ethyl mercuric bromide included the following absorptions with the probable assignments: (cm^{-1}) 297 m; 253 m (?methyl torsion); 207 m (Hg–Br stretch); 154 sh; 70, 35 s, br (C–Hg–Br def.).

E. Benzyl Mercuric Halides

Chloride and bromide compounds were examined in the infrared to 200 cm^{-1} and spectra were interpreted on the basis of probable assignments for benzyl iodide; low-frequency results for these, and for dibenzyl mercury, are shown in Table IX. The notation for the fundamentals, except for a change in axes, is that of Whiffen (1956).

TABLE IX. Low-frequency Absorptions and Modes in Benzyl Mercury Compounds (cm^{-1}).

$(C_6H_5CH_2)_2Hg$	$(C_6H_5CH_2)HgCl$	$(C_6H_5CH_2)HgBr$	Mode
343 w	328 w	319	*u*X-sensitive
	307 s		Hg–Cl stretch
220 m	292 sh	296 m 291 sh	benzyl–Hg def.
		226 s	Hg–Br stretch

TABLE X. Infrared Frequencies (cm^{-1}) and Assignments for Diphenyl Mercury, Phenylmercuric Chloride, Bromide and Iodide.

$(C_6H_5)_2Hg$	C_6H_5HgCl	C_6H_5HgBr	C_6H_5HgI	Mode
	330 vs			Hg···Cl str.
	310 sh	312 w	303 w 292 sh	combination?
		257 s		Hg···Br str.
258 s 252 s 248 sh	220 m	216 w	242 m 216 vw	*t*X-sens. (a_1)
207 m	184 sh	198 m	208 m 181 sh	*u*X-sens. (b_2)
			170 s	Hg···I str.
100 m	180 s	175 s, br.	160 m	*x*X-sens. (b_1)
	100 sh 78 s	~95 w, sh 59 s	87 w, br. 51 s	(C_6H_5···Hg···Hal) def.

F. Phenyl Mercuric Halides

The region 375–40 cm^{-1} has been considered separately for diphenyl mercury, and phenyl mercury chloride, iodide and bromide; data are given in Table X.

VII. Dimethyl Zinc

This molecule, the simplest organo-zinc compound, has been investigated in the Raman and infrared by, *inter alia*, Thompson *et al.* (1940) and Gutowsky (1949). It represents an interesting example of a structure determination by vibrational spectroscopy, since earlier electron diffraction studies by Brockway and Jenkins (1936) were unable to confirm the expected linearity because of the high background of atomic scattering. The possibility of a bent skeleton was dismissed, as the C_{2v} symmetry predicts twenty-one fundamentals, all active both in the infrared and the Raman, and the observed number of coincidences is clearly very much less than this. There are three linear structures: with staggered methyl groups and D_{3d} symmetry, with eclipsed methyl groups and D_{3d} symmetry, and with free rotation of the methyl groups and D_{3h} symmetry. The selection rules are discussed for all possibilities by Thompson (*vide supra*) and also Bauman (1956). The situation is entirely similar to that of dimethyl mercury (see Section VIA), and the assignments for the low-frequency mode ν_{11}, recommended by Gutowsky following a normal coordinate analysis, are shown in Table XI. The linear skeleton with D_{3h} symmetry and free rotation of the methyl groups is favoured.

TABLE XI. Low-frequency Assignments and Force Constants in Dimethyl Zinc

Species	Mode	Force Constant md.Å^{-1}	cm^{-1}
ν_3	C–Zn–C stretch	2.39	504
ν_{11}	C–Zn–C bend	0.09	144

VIII. Tin Compounds

Considerable low-frequency data, Raman and infrared, are available for tin-containing molecules, both with relatively simple systems for which complete vibrational assignments are given and also, in less formal fashion, for complex molecules and derived addition compounds (see, for example, Poller 1966). Bentley *et al.* (1968) have reviewed reported Sn–C stretching frequencies (about 500–570 cm^{-1}) in alkyl compounds, and provide a list of relevant references; they also refer to various modes in phenyl-tin compounds.

A. Tetramethyl Tin

This tetrahedral (point group T_d), symmetrical, molecule has been extensively studied by Edgell and Ward (1962) and Lippincott and Tobin (1953), and was completely assigned before low-frequency infrared data were available. The two Raman-active skeletal C–Sn–C bending modes are the only low-frequency fundamentals and are placed near 160 and 145 cm^{-1} (Lippincott and Tobin 1953). It is agreed that there is little coupling between methyl groups, and the given descriptions of vibrations in terms of stretching, bending, and so on, modes are satisfactory; thermodynamic functions are also tabulated.

B. Tetraethyl and Tetra (*n*-propyl) Tin

A thorough far-infrared investigation to 55 cm^{-1} and discussion by Wood and Taimsalu (1963) has enabled complete assignments and thermodynamic functions to be made. Assignments were based on measurements in higher-frequency ranges and also by reference to similar molecules, such as tetramethyl tin (*vide supra*) and *n*-propyl halides (Brown and Sheppard 1954). Some results are shown in Table XII.

TABLE XII. Low-frequency Infrared Assignments for Tetraethyl and Tetra (*n*-propyl) tin (cm^{-1}).

<table>
<tr><th rowspan="2">Mode</th><th rowspan="2">$SnEt_4$</th><th colspan="2">Sn(*n*-Pr)$_4$</th></tr>
<tr><th>*trans*</th><th>*gauche*</th></tr>
<tr><td>C–C–C bend</td><td></td><td>?294 vs</td><td>390 s</td></tr>
<tr><td>Sn–C–C bend</td><td>272 s</td><td>225 m</td><td>294 vs</td></tr>
<tr><td>C–Sn–C bend</td><td>132 m
86 vw</td><td colspan="2">129 s</td></tr>
<tr><td>C–C torsion</td><td></td><td colspan="2">97 s</td></tr>
<tr><td>Sn–C torsion</td><td>(33)</td><td></td><td></td></tr>
</table>

The observed low-frequency spectrum of $SnEt_4$ was simple, and no effects due to torsional isomerism of the ethyl groups were noted; as with $SnMe_4$ it was concluded that there was little coupling between vibrations of alkyl groups, and that no change in simple valence field force constants derived from $SnMe_4$ are needed.

An interesting complication with the *n*-propyl compound is the possibility of *trans* and *gauche* conformations of the propyl group with respect to orientation about the C_1C_2 bond axis. By analogy with the spectrum of *n*-propyl halides above 500 cm^{-1}, it was concluded that both forms were present in substantial amounts. The strong 97 cm^{-1} absorption was assigned as a torsion and an approximate calculation of the barrier height, V_3, based on a harmonic oscillator fitted to a cosine potential, using $V_3 = (8/9)\pi^2\nu^2 I_{red}$, with $I_{red} = 21.3$ amu Å^2, gave a value of 3.8 kcal $mole^{-1}$. The torsional frequency about the Sn–C bond in tetraethyl tin was calculated as 33 cm^{-1}, which was below the range of measurements.

C. Methyl Stannanes

This is one of the few examples in which infrared spectra of isotopically differing molecules have been measured in the far infrared. Kimmel and Dillard (1968) have reported spectra to 250 cm^{-1} for methyl stannane, methyl stannane-d_3 and methyl-d_3 stannane. Low-frequency results are shown in Table XIII, the notation being that of Herzberg (1945).

TABLE XIII. Low-frequency Infrared Assignments for Methyl Stannanes (cm^{-1}).

Species	Mode	CH_3SnH_3	CH_3SnD_3	CD_3SnH_3
$\nu_{12}(e)$	SnH_3 rock	416.3 m	316.6 m	392.4 w
?			390 w	
?		340 vw		
$\nu_5 - 2\nu_6(a_1)$		320 vw		
$\nu_{12} - \nu_6(e)$		298 vw		
ν_6	torsion	(109)	(101)	(88)

Methyl stannane belongs to the point group C_{3v}, with five non-degenerate a_1 species, six degenerate e, and one a_2; all except the last are infrared-active, and since most of the spectrum is available in the higher-frequency range, complete assignment is possible. Microwave studies on methyl stannane (Cahill and Butcher 1961) have established a barrier to rotation of 650 cal $mole^{-1}$, from which a value of 109 cm^{-1} for the infrared-inactive torsional frequency ν_6 was calculated. The combination band $\nu_4 + 2\nu_6$ at 910 cm^{-1} (not shown) and the above combinations $\nu_5 - 2\nu_6$ at 320 cm^{-1} and $\nu_{12} - \nu_6$ at 298 cm^{-1} gives values 108, 104 and 118 cm^{-1}, respectively, for this mode. Assuming the same barrier for the deuterated stannanes, the torsional frequencies are as above.

A normal coordinate calculation using anharmonicity constants of the C–H stretch and CH_3 antisymmetric deformations transferred from methane was made, and force constants determined. The potential energy distribution is listed and it is concluded that, except for the

SnH_3 deformation and the C–Sn stretching modes of the a_1 symmetry species, the concept of characteristic group frequencies is valid for these compounds. For the *e* symmetry species it was shown that the CH_3 rocking, SnH_3 deformation, and SnH_3 rocking modes show a significant amount of mixing for methyl and methyl-*d* stannane, but the CH_3 rocking vibration in methyl stannane-*d* is a pure mode. Similar treatment of methyl silane and methyl germane show that the degree of localization of this mode increases with deuteration of the SiH_3 and GeH_3 groups. This suggests that if the hydrogen atom on the MH_3 group of a methyl-MH_3 molecule is replaced by heavier atoms, then the CH_3 rocking mode will become independent of the rest of the molecule and function as a characteristic group vibration. Evidence for this effect in the methyl chloro-silanes and -germanes has been given in literature, and Edgell and Ward (1962) and Markova (1966) are cited as examples.

D. Phenyl Tin Compounds

Recent vibrational studies, including low-frequency Raman and infrared measurements, are available for triphenyl tin halides, diphenyl tin dihalides, tetraphenyl tin and miscellaneous organo-tin compounds (Poller, 1966), and phenyl tin trichloride (Durig *et al.* 1966). A review of these and other reports, together with additional measurements, and correlations with analogous group IV A compounds, is given by Smith (1968).

1. *Phenyl Tin Trichloride*

A detailed infrared and Raman study by Durig *et al.* (1966) has been reported, with measurements to 33 cm^{-1} for trichlorophenyl-silane, -germane and -stannane. Very recently a far-infrared and Raman study of pentadeuterophenyltrichlorogermane (Durig and Sink 1968) has been reported.

The possibility of Raman lines at 86, 75 and 63 cm^{-1} in the three silicon, germanium and tin compounds, respectively, arising from torsional vibrations was considered. Approximate calculations of barrier heights gave values of 2521, 1984 and 1414 cal. $mole^{-1}$. Since the largest unambiguous six-fold barrier previously recorded by Fateley and Miller (1963) was about 50 cal $mole^{-1}$ this assignment was rejected. A similar calculation for the 118 cm^{-1} Raman line found by Stephenson and Coburn (1965) in phenyl trichloromethane gave a barrier of similar order, about 4.5 kcal. Hence, it was assumed that the lowest Raman lines arise from the phenyl $\nu_{20'}$ mode, and that the torsional frequency has not been detected either in the infrared or the Raman, and probably lies below the limit of measurement (33 cm^{-1}).

2. *Tetraphenyl Tin*

No detailed analysis of this has appeared, but low-frequency infrared data are available from, *inter alia*, Smith (to 50 cm^{-1}) and Poller, *vide supra* (to 250 cm^{-1}). Assignments due to Smith for this and other group IV A molecules are collected in Table XIV. However, the observed frequencies are not entirely consistent with those of Poller, and there are also differences in assignment from those of Brown *et al.* (1965), so no definitive list can be given. The notation is that of Whiffen (1956). Splitting arises, especially with the *t* mode when more than one phenyl group is attached to the central atom.

TABLE XIV. Low-frequency Infrared Assignments in Tetraphenyl Group IV A Compounds (cm^{-1}).

Species	$(C_6H_5)_4C$	$(C_6H_5)_4Si$	$(C_6H_5)_4Ge$	$(C_6H_5)_4Sn$	$(C_6H_5)_4Pb$
t	635 s	435 m	332 m	268 s	224 s
y	527 m	519 m	481 m	459 m	453 m
y'	492 m	511 vs	465 m	448 m	442 m
			269 m		
w		399 vw	399 vw	399 vw	396 vw
w'	395	390 vw	390 vw	391 vw	389 vw
u	332 w	261 w	232 m	225 vw	
u'	292 w	223 w	214 w		
t'	242 w	239 m		212 m	205 m
x	~200 vw	185 w	187 w	193 m	183 ms
x'		171 m	168 m	152 mw	148 m
δ_f	96 vw	98 w	98 w	89 w	89 w
δ_e	63 vw	57 vw	69 vvw	61 w	55 w

E. Miscellaneous Tin Compounds

1. *Cyclic Compounds*

Sulphur-containing cyclic organo-tin compounds of formula I and II

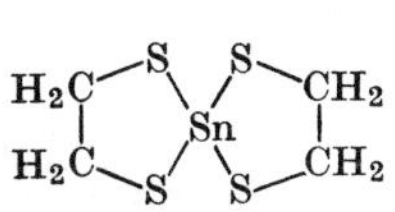

I

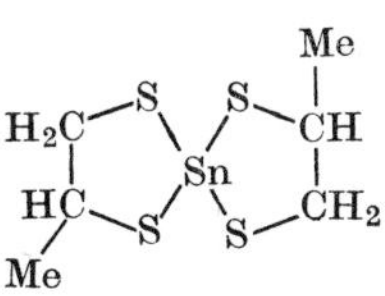

II

I. 1,4,6,9-tetrathia-5-stannaspiro [4,4] nonane.

II. 3,7-dimethyl-1,4,6,9-tetrathia-5-stannaspiro [4,4] nonane.

have been examined in the crystalline state to 110 cm^{-1} by Finch *et al.* (1965). Sulphur-tin stretching vibrations were found to vary from 392–330 cm^{-1} (symmetric) and 363–267 cm^{-1} (antisymmetric). Interpretation on the basis of planar rings, giving V_d symmetry, and considering the skeletal modes independently of the CH modes, gave satisfactory results.

2. *Triphenyl Tin Fluoride and Hexaphenyl di-tin*

Low-frequency infrared spectra to 250 cm^{-1} have been reported by Poller (*vide supra*). Triphenyl-tin fluoride was shown to be quite unlike the analogous chlorine, bromine and iodine compounds, presumably due to its inferred polymeric, fluorine-bridged structure; it is the only triphenyl-tin compound having a single sharp absorption in the 440 cm^{-1} region.

IX. Arsenic Compounds

A. Methyl Arsine

There have been several investigations of the vibrational spectrum of this compound and of related CH_3MX_2 molecules. Harvey and Wilson (1966b) have reported Raman and infrared spectra of CH_3AsH_2, CH_3AsD_2, CD_3AsH_2 and CD_3AsD_2, using grating and interferometric instruments to cover the far-infrared region to <200 cm^{-1}. Crystal, liquid, and gas-phase samples were examined, a normal coordinate calculation was made, and a complete vibrational assignment proposed. As in the somewhat analogous examples of methylselenol (Harvey and Wilson 1966a), the geometry of the molecule was inferred from the known structure of analogues (methyl amine and methyl phosphine), and C_s symmetry presumed. For this molecule, there should then be fifteen normal modes, all active in both the Raman and the infrared.

In the low-frequency region two points of interest arise, the torsional vibration, and the possibility of resolving disputed assignments in related molecules by considerations developed as a result of the intensive study of methyl arsine. The torsional oscillation of the AsH_2 group against the methyl moiety belongs to symmetry class a'' (which contains six modes) and is expected to be active in both the Raman and the far-infrared, although there is no evidence of any great intensity. The far-infrared spectrum of methyl arsine at 1 atmosphere pressure in a 1-m cell shows a very weak, broad absorption centred near 185 cm^{-1}; as with methylselenol the lack of definition is presumably due to the inherent weakness of the band and to overlapping with hot bands. A calculation of the barrier height gives a value of about 500 cm^{-1} with no error limits quoted; this is certainly of the right order of magnitude, and Harvey and Wilson (*vide supra*) quote a value of 463 cm^{-1} for the corresponding difluoro-compound, CH_3AsF_2, derived by Nugent and Cornwell (1962) from microwave studies.

Previous assignments for various deformation modes in CH_3MX_2 molecules have been at variance. Thus, in CH_3PCl_2, low frequency (Raman) absorptions at 281 and 193 cm^{-1}, which have been assigned to the wagging and scissoring modes, respectively, by Griffiths (1965) have been placed in the reverse order by Durig *et al.* (1965). Also, the twisting mode in methylamine was disputed and uncertain in methyl phosphine. By arguments developed from assignments in methyl arsine, and related molecules, Harvey and Wilson recommend the assignments shown in Table XV.

TABLE XV. Some Vibrational Assignments in CH_3MX_2 Molecules (cm^{-1}).

Scissoring of MX_2 group		Wagging and Twisting of MX_2 group
CH_3PCl_2	193	281,238
CH_3PH_2	1,093	740 (700)
CH_3AsH_2	973	674,654
CH_3NH_2	1,623	780 (720)

B. Triethynyl Arsenic Compounds

The very interesting compounds M(C≡CH)$_3$, where M is phosphorus, arsenic or bismuth were first made in 1964 by Voskuil and Arens, and have been subjected to a thorough infrared (to 35 cm^{-1}) and Raman study in the vapour and solution phases by Miller *et al.* (1967a, b); the antimony compound, which had very low vapour pressure, was also examined at 100°K as a condensed film. A vibrational analysis located thirty-eight out of the forty-two fundamentals expected on the basis of presumed trigonal pyramidal structures, and thus confirmed the assumed C_{3v} structure. Low-frequency data are shown in Table XVI.

TABLE XVI. Low-frequency Assignments in Some Group V Triethynyls (cm^{-1}).

Mode	P(C≡CH)$_3$	As(C≡CH)$_3$	Sb(C≡CH)$_3$
MC_3 def. ν_6	120	89	[72?]
M–C≡C bend ν_{14}	423.5	378.5	325
M–C≡C bend ν_{15}	266	231	219
MC_3 def. ν_{16}	100.5	100	94

Lack of sample prevented the running of solution spectra of the antimony compound below 400 cm^{-1} in the infrared, and the ν_6 mode is uncertain; the authors point out that their assignments put $\nu_6 < \nu_{16}$ for the arsenic compound, but $\nu_6 > \nu_{16}$ for the phosphorus compound. A similar study has also been made for the group IV tetraethynyls of silicon, germanium and tin, which are apparently considerably less stable and are more dangerously explosive.

X. Phenyl Arsenic and Antimony Compounds

A recent paper by MacKay *et al.* (1968) reports infrared spectra to 250 cm^{-1} for many phenyl and phenylhalogen compounds of germanium, arsenic and antimony. In addition, an extensive listing of relevant literature is given and, as a result of correlation between their own and reported spectra for series such as $(C_6H_5)_nMCl_{4-n}$, with M = Si, Ge, Sn, assignments of ring X-sensitive modes and metal-chlorine vibrations are made. Some of these differ, as was noted earlier, from assignments for germanium compounds by Durig and tin compounds by Poller; the original paper may be consulted for details. Pentaphenyl arsenic and antimony compounds, the structures of which in the crystal phase are known to be, respectively, trigonal bipyramidal and square pyramidal (Wheatley 1964a, b), were examined together with various halogenated derivatives of general formula $(C_6H_5)_nMX_{5-n}$, with M = As and Sb and X = Cl, Br, I. A summary of the proposed phenyl assignments using Whiffen's (1956) nomenclature is given in Table XVII which also includes analogous phosphorus compounds.

The authors point out that the limited spectroscopic evidence available

TABLE XVII. Low-frequency Phenyl Modes in Group V Phenyl Compounds (cm^{-1}).

	$(C_6H_5)_3P$	$(C_6H_5)_4PX$	$(C_6H_5)_5P$	$(C_6H_5)_3As$	$(C_6H_5)_4AsX$
q	1089	1106	1085	1072	1080
y	520	527	535	471	455–480
t	428	446	440	311	344–364

	$(C_6H_5)_5As$	$(C_6H_5)_3Sb$	$(C_6H_5)_4SbX$	$(C_6H_5)_5Sb$
q	1060	1063	1054–1067	1056
y	480	453	440–460	451
t	320	268	260–290	266

does not allow a definite assignment of pentaphenyl antimony to the square pyramidal structure, but discuss differences between the spectrum of this, and of $(C_6H_5)_5As$. In all the examples the absorption assigned to the y mode appears as a complex series apparently dependent on the local site symmetry. Further low-frequency determinations and Raman measurements are necessary in order to confirm and extend interpretations.

XI. Mono- and Di-substituted Benzene Compounds

A particularly important paper is the recent report by Griffiths and Thompson (1967) of a detailed far-infrared interferometric study of a large number of mono- and di-substituted benzenes from 450–50 cm^{-1}. Compounds were measured as liquids, in solutions, and in polyethylene matrices and as low-temperature solids.

The mono-substituted compounds C_6H_5X($X = F$, Cl, Br, I, CH_3) were considered and identification of the lowest frequency fundamentals established. Using Whiffen's (1956) notation, the following correlation diagram, redrawn from the figures quoted in Table XVIII is given as Fig. 2; a more complete diagram is available in the original reference.

TABLE XVIII. Low-frequency Fundamentals in Mono-Substituted Benzenes (cm^{-1}).

Compound	Bands observed	Values adopted as fundamentals
Fluorobenzene	242, 405	242, 405, 499, 519, 610
Chlorobenzene	196, 298, 420	196, 298, 416, 467, 610
Bromobenzene	180, c. 260, 316	180, 260, 314, 455, 610
Iodobenzene	167, 221, 267, 421	167, 220, 267, 448, 610
Toluene	217, 346, 428?	217, 347, 465, 521, 615

Di-halobenzenes: these have been the subject of many investigations. The *para*-halofluorobenzenes FC_6H_4X ($X = Cl$, Br, I) have been studied independently by, *inter alia*, two groups, Narasimham and Nielsen (1956) and Griffiths and Thompson (*vide supra*), and some assignments differ. These molecules have C_{2v} symmetry and the thirty fundamentals divide into the species: $11a_1 + 10b_2$ (in plane) and $3a_2 + 6b_2$ (out of plane), the a_2 species being active in the infrared. However, Ingold *et al.* (1946) showed that in monodeuterobenzene only that a_2 mode which corresponds to the e_{1u} fundamental in benzene gives an actual change in

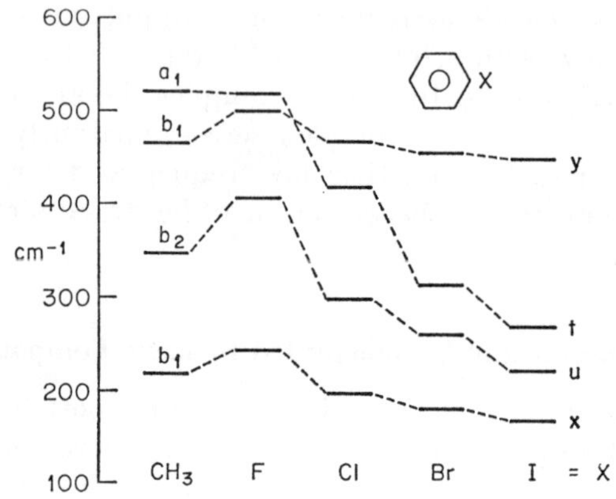

FIG. 2. Correlation chart showing some low-frequency assignments in mono-substituted benzene compounds.

polarizability; the two a_2 modes which correspond to the benzene e_{2u} fundamentals leave all components of the polarizability unaltered. Thus, only one of the a_2 modes might be expected to show appreciable intensity in the Raman. The assignments of Narasimham and Nielsen are shown in Table XIX.

TABLE XIX. Assignments of Low-frequency Modes in Para-halofluorobenzenes (Narasimham and Nielsen) (cm^{-1}).

Species	Fluorochloro-	Fluorobromo-	Fluoroiodo-benzene	Mode
a_1	376	290	244	C–C–C bend
b_1	336	323	309	C–X bend
b_1	420	416	411	C–F bend
b_2	267	224	196	C–X bend
a_2	(368)	340	(340)	C–C–C bend

It will be noted that the lowest assignment is 196 cm^{-1}. Griffiths and Thompson, however, assign the lowest absorption for each compound to a b_1 mode, all substantially lower than the previous assignments. The next lowest assignment is ascribed to an in-plane bending mode, b_2, again differing from those of Narasimham and co-workers. These and other low-lying assignments are collected in Table XX.

A correlation diagram showing both assignments is shown in Fig. 3. In addition to the *para*-fluorohalobenzenes, spectra and assignments are reported for *para*-dichlorobenzene, *para*-dibromobenzene, *para*-diiodobenzene and *para*-bromoiodobenzene. It is noted that the ring-breathing

TABLE XX. Assignments of Low-frequency Modes in Para-halofluorobenzenes (Griffiths and Thompson) (cm^{-1}).

	b_1	b_2	a_1	b_1	b_2
p-Difluorobenzene	166(b_{3u})	350(b_{2u})	451R(a_g)	375R(b_{2g})	430, solid (b_{3g})
p-chlorofluorobenzene	137	266	374, solid	336	421
p-bromofluorobenzene	124	224	290	323	420
p-iodofluorobenzene	114	198	245	314	416

Raman results from Whiffen and Stojiljkovic (1958a, b).

mode (a_1) appears to be most sensitive to the mass of substituents, and also that there is a definite shift to higher frequencies of 10–20 cm^{-1} for the lowest frequency below 160 cm^{-1} for a crystalline solid at $-170°$ compared with the absorption at 20° in the liquid phase.

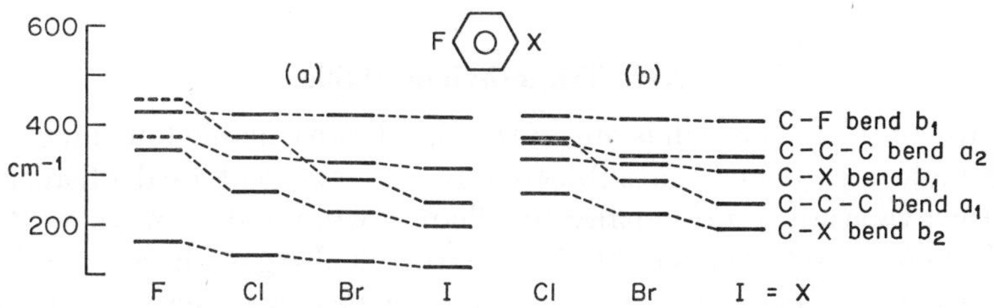

FIG. 3. Correlation charts showing some low-frequency assignments in di-substituted benzene compounds, (a) from Griffiths and Thompson, (b) from Narasimham and Nielsen.

Griffiths and Thompson have also given a rare example of quantitative intensity measurements in this region of the spectrum. Using benzene solutions in a 3-mm path-length cell, they obtained values of the integrated intensity A, and the extinction coefficient ϵ [defined by the expressions $A = (1/cl) \int \ln(I_0/I) d\nu$ and $\epsilon = (1/cl) \log_{10}(I_0/I_{max})$ using conventional symbols] and employing several slit-widths. Results, accurate to within 10%, are shown in Table XXI.

It was found that the intensity of this band, corresponding to the lowest frequency (in the b_1 or b_{2u} class) varied with X and Y more than that of other bands. If, as a crude approximation, the molecules are considered as linear systems, X–Y–Z, where X and Z are halogens, and Y is a mass of 76, then it is possible to show that the bending frequency is inversely proportional to the reduced mass of the system, m. Plots of A *versus* $1/m$ are presented and shown to be linear, and this is interpreted as further evidence for the correct assignment of the band concerned

TABLE XXI. Intensity Measurements for b_1 (x-type) Band in *Para*-dihalobenzenes.

	Peak frequency (cm^{-1})	ϵ	A (10^3 cm mole^{-1})
Difluoro	167	7.54	205
Fluorochloro	138	2.53	103
Fluorobromo	125	1.05	57.5
Fluoroiodo	114	0.48	29.4
Dichloro	116	0.92	52.6
Chlorobromo	104	0.35	20.1
Chloroiodo	92	0.16	—

to the out-of-plane bending mode. *Para*-substituted toluenes and nitrobenzenes were also studied in the far-infrared, and assignments and correlation charts are given. An infrared and Raman investigation of *ortho*-fluorochlorobenzene by Narasimham and Nielsen (1956) gives a complete vibrational assignment of this molecule.

XII. Trimethylene Oxide

This compound, which is one of the simplest available four-membered ring molecules, has been subjected to a very great deal of accurate microwave studies, principally by Chan, Gwinn and co-workers (Chan *et al.* 1966 and references quoted therein), and also gas-phase far-infrared investigations, especially by Lord and colleagues (see, for example, Durig and Lord 1966). The main point of interest is the ring-puckering vibration, which is active in the infrared, and the form of the potential function governing this mode. Both microwave and far-infrared techniques were found to be necessary.

Values for pure rotational transitions of several vibrational levels of the puckering mode were evaluated from microwave studies. The proposal was made that the potential function was of double-minimum type, with a small barrier later set at 15 cm^{-1} at the planar conformation, as shown in Fig. 4, in order to explain the fact that the (observed) rotational constants, when plotted against the vibrational quantum number of the puckering mode, did not give the usual linear relation, but exhibited zigzagging and curvature. Far-infrared bands at 89.8, 118.3, 128.9, 139.0, and so on, and later 105.2 cm^{-1} had been observed and assigned to the vibrational states of the puckering mode. From considerations arising from the microwave spectra it was necessary to assume that the 118.3 cm^{-1} absorption represented the 1–2 transition, which led to the prediction of the 0–1 transition at 70 ± 5 cm^{-1}. This was difficult on position and intensity grounds to reconcile with the occur-

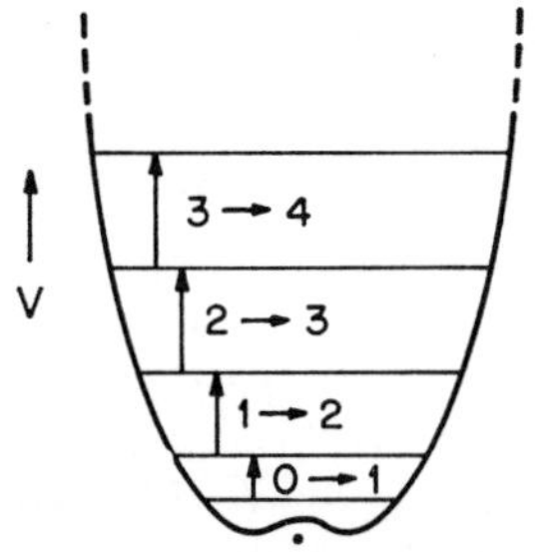

FIG. 4. Form of ring-puckering potential in trimethylene oxide.

rence of the 89.8 and the 105.2 cm^{-1} bands. A re-examination of the far-infrared spectrum was, therefore, undertaken and although the previous absorptions were all confirmed, a new band at 53.4 cm^{-1} was noted, and assignment of this to the $0 \rightarrow 1$ transition was made. This new assignment was in agreement with a value of 60 ± 20 cm^{-1} obtained from microwave intensity measurements of the two strongest vibrational satellites for several rotational transitions. The potential function was then calculated, and used to predict the far-infrared spectrum of the deuterated molecule. As shown in Table XXII, the agreement between observed and calculated results is excellent. The energy levels for this molecule were computed using the modified quartic function: $V(X) = 8.664(X^4 - 2.6571X^2)$ cm^{-1}. An important point is that in such studies, and in any investigations in which hot bands are involved, gas-phase measurements are essential.

TABLE XXII. Far Infrared Transitions in Trimethylene and d-Trimethylene Oxide (TMO) (cm^{-1}).

Transition	TMO	d_6-TMO (observed)	d-TMO (calculated)
0–1	$53.4 \pm .2$[a]	—[b]	38.7
1–2	$89.9 \pm .2$	69.3	68.6
2–3	$104.7 \pm .2$	80.7	79.9
3–4	$118.1 \pm .5$	91.0	90.5
4–5	$128.8 \pm .5$	99.0	99.0
5–6	$138.9 \pm .5$	106.4	106.4
6–7	$147.3 \pm .5$	112.6	113.0
7–8	$155.0 \pm .5$	118.0	118.9
8–9	$161.6 \pm .5$	124.0	124.4

[a] Chan *et al.* (1966). [b] Lafferty and Lord, quoted on p. 1108 of Chan *et al.* (1966).

References

Bauman, R. P. (1956). *J. Chem. Phys.* **24**, 13.
Bell, R. P. (1945). *Proc. R. Soc.* (A), **183**, 328.
Bellamy, L. J. (1962). "Infrared Spectra of Complex Molecules." Wiley, New York.
Bellamy, L. J. (1968). "Advances in Infrared Group Frequencies." Methuen, London.
Bentley, F. F., Smithson, L. D., and Rozek, A. L. (1968). "Infrared Spectra and Characteristic Frequencies 700 to 300 cm^{-1}." Interscience, New York.
Brockway, L. O. and Davidson, N. R. (1941). *J. Am. chem. Soc.* **63**, 3287.
Brockway, L. O. and Jenkins, H. O. (1936). *J. Am. chem. Soc.* **58**, 2036.
Brown, J. K. and Sheppard, N. (1954). *Spectrochim. Acta* **50**, 1164.
Brown, D. H., Mohammed, A., and Sharp, D. W. A. (1965). *Trans. Faraday Soc.* **21**, 659.
Cahill, P. and Butcher, S. (1961). *J. Chem. Phys.* **35**, 2255.
Chan, S. I., Borgers, T. R., Russell, J. W., Strauss, H. L., and Gwinn, W. D. (1966). *J. Chem. Phys.* **44**, 1103.
Daykin, P-N, Sundaram, S., and Cleveland, F. F. (1962). *J. Chem. Phys.* **37**, 1087.
Durig, J. R. and Lord, R. C. (1966). *J. Chem. Phys.* **45**, 61.
Durig, J. R. and Sink, C. W. (1968). *Spectrochim. Acta* **24**, 575.
Durig, J. R., Block, F., and Levin, I. W. (1965). *Spectrochim. Acta* **21**, 1105.
Durig, J. R., Sink, C. W., and Bush, S. F. (1966). *J. Chem. Phys.* **45**, 66.
Edgell, W. F. and Ward, C. H. (1962). *J. molec. Spectrosc.* **8**, 343.
Fateley, W. G. and Miller, F. A. (1963). *Spectrochim. Acta* **19**, 611.
Finch, A., Poller, R. C., and Steele, D. (1965). *Trans. Faraday Soc.* **61**, 2628.
Gray, A. P. (1963). *Can. J. Chem.* **41**, 1511.
Green, J. H. S. (1968). *Spectrochim. Acta* **24**, 863.
Griffiths, J. E. (1965). *Spectrochim. Acta* **21**, 1135.
Griffiths, P. R. and Thompson, H. W. (1967). *Proc. Roy, Soc.* (A), **298**, 51.
Groenewege, M. P. (1958). *Z. Phys. chem.* **18**, 147.
Groenewege, M. P. (1959). *Rev. Univlle Mines*, 9 Series, **15**, 461.
Gutowsky, H. S. (1949). *J. Chem. Phys.* **17**, 128.
Hall, J. R. and Mills, J. C. (1966). *J. Organometal. Chem.* **6**, 445.
Harvey, A. B. and Wilson, M. K. (1966a). *J. Chem. Phys.* **45**, 678.
Harvey, A. B. and Wilson, M. K. (1966b). *J. Chem. Phys.* **44**, 3535.
Herzberg, G. (1945). "Infrared and Raman Spectra of Polyatomic Molecules," Vol. 2. Van Nostrand, New York.
Hidalgo, A. (1958). *C. r. hebd. Seanc. Acad. Sci., Paris* **249**, 395.
Hidalgo, A. (1962). *An. R. Soc. esp. Fis. Quim.* **58(A)**, 71.
Hoffmann, E. G. (1960). *Z. Electrochem.* **64**, 616.
Ingold, C., Bailey, C. R., Gordon, R. R., Hale, J. B., Herzfeld, N., and Poole, H. G. (1946). *J. Chem. Soc.* 299.
Jackson, J. A. and Nielsen, J. R. (1964). *J. molec. Spectrosc.* **14**, 320.
Jones, W. J. and Sheppard, N. (1965). quoted by Scott *et al.* in (1965) *J. Phys. Chem.* **63**, 1139.
Karapinka, G. L., Smith, J. J., and Carrick, W. L. (1961). *J. Polym. Sci.* **50**, 143.
Kimmel, H. and Dillard, C. R. (1968). *Spectrochim. Acta* **24**, 909.
Lippincott, E. R. and Tobin, M. C. (1953). *J. Am. chem. Soc.* **75**, 4141.
Lord, R. C. and Miller, F. A. (1956). *Appl. Spectrosc.* **10**, 115.
MacKay, K. M., Sowerby, D. B., and Young, W. C. (1968). *Spectrochim. Acta* **24**, 611.

Markova, S. V. (1966). "Supplement on Molecular Spectroscopy", p. 88. Optical Society of America.
Miller, F. A. and Lemmon, D. H. (1967a). *Spectrochim. Acta* **23**, 1099.
Miller, F. A., Sacher, R. E., and Lemmon, D. H. (1967b). *Spectrochim. Acta* **23**, 1169.
Narasimham, N. A. and Nielsen, J. R. (1956). *J. Chem. Phys.* **24**, 433.
Nugent, L. J. and Cornwell, C. D. (1962). *J. Chem. Phys.* **37**, 523.
Nyquist, R. A. and Potts, W. J. (1960). *Spectrochim. Acta* **16**, 419.
Ogawa, T. and Miyazawa, T. (1964). *Spectrochim. Acta* **20**, 557.
Ogawa, T., Hirota, K. and Miyazawa, T. (1965). *Bull. chem. Soc. Jap.* **38**, 1105.
Poller, R. C. (1966). *Spectrochim. Acta* **22**, 935.
Rundle, R. E. and Lewis, P. H. (1953). *J. Chem. Phys.* **21**, 986.
Scott, D. W., Good, W. D., Lacina, J., and McCullough, J. P. (1959). *J. Phys. Chem.* **63**, 1139.
Scott, D. W., Guthrie, G. B., Messerly, J. F., Todd, S. S., Berg, W. T., Hossenlopp, I. A., and McCullough, J. P. (1962). *J. Phys. Chem.* **66**, 911.
Scott, D. W., Crowder, G. A., Gorin, G., and Kruse, F. H. (1965). *J. molec. Spectrosc.* **16**, 115.
Sheline, R. K. and Pitzer, K. S. (1950). *J. Chem. Phys.* **18**, 595.
Smith, A. L. (1968). *Spectrochim. Acta* **24**, 695.
Stephenson, C. V. and Coburn, W. C. (1965). *J. Chem. Phys.* **42**, 35.
Stewart, J. E. (1965). *In* "Interpretive Spectroscopy." (Freeman, K. ed.) p. 137 ff. Reinhold. New York.
Thompson, H. W., Linnett, J. W., and Wagstaffe, F. J. (1940). *Trans. Faraday Soc.* **36**, 797.
van der Kelen, G. P. and Herman, M. A. (1956). *Bull. Soc. chem. Belges.* **65**, 362.
Voskuil, W. and Arens, J. F. (1964). *Rec. Trav. chim. Pays-Bas Belg.* **83**, 1301.
Wheatley, P. J. (1964a). *J. Chem. Soc.* 2206.
Wheatley, P. J. (1964b). *J. Chem. Soc.* 3718.
Whiffen, D. H. (1956). *J. Chem. Soc.* 1350.
Whiffen, D. H. and Stojiljkovic, A. (1958a). *Spectrochim. Acta* **12**, 47.
Whiffen, D. H. and Stojiljkovic, A. (1958b). *Spectrochim. Acta* **12**, 57.
Wilson, E. B. and Thomas, C. (1966). *quoted in J. Chem. Phys.* **44**, 3535, paper by A. B. Harvey and M. K. Wilson.
Wood, J. L. and Taimsalu, P. (1963). *Trans. Faraday Soc.* **59**, 1754.
Woodward, L. A. (1963). *Spectrochim. Acta* **19**, 1963.

Chapter 8

SPECTROSCOPY OF CRYSTAL LATTICES

This chapter is concerned with the interaction of infrared radiation with the various modes of crystal lattices. Infrared spectroscopy is capable of providing information concerning the vibrational frequencies of a crystal and the interatomic forces operative within the lattice. The various forms of absorption (that is, Reststrahlen, multiphonon, defect-induced) observed with the different types of crystal lattice (that is, ionic, covalent, defect, molecular) are treated in the theoretical section and characteristic examples are provided in the section on applications. The vast majority of the results was obtained from infrared measurements of one form or another (transmission, reflection, absorption) but due to the complementary nature of Raman, and especially slow-neutron spectroscopy, certain results quoted were obtained using these methods.

I. Theoretical Considerations

A. Vibrations of Crystal Lattices

Since in many crystals, individual chains of molecules can be distinguished where the interactions between molecules in a given chain are considerably stronger than between molecules in different chains, it is not surprising that many of the features of a three-dimensional crystal are merely extensions of the one-dimensional model. Initially, a linear diatomic chain with alternating masses M and m, separated by a distance d, situated on lattice sites $2j$ and $2j+1$ respectively, and, having an interparticle force constant f will be considered. The equations of motion, involving the displacements from equilibrium, u_{2j} and u_{2j+1}, are

$$M\frac{d^2 u_{2j}}{dt^2} = f(u_{2j+1} + u_{2j-1} - 2u_{2j})$$

and

$$m\frac{d^2 u_{2j+1}}{dt^2} = f(u_{2j+2} + u_{2j} - 2u_{2j+1}).$$

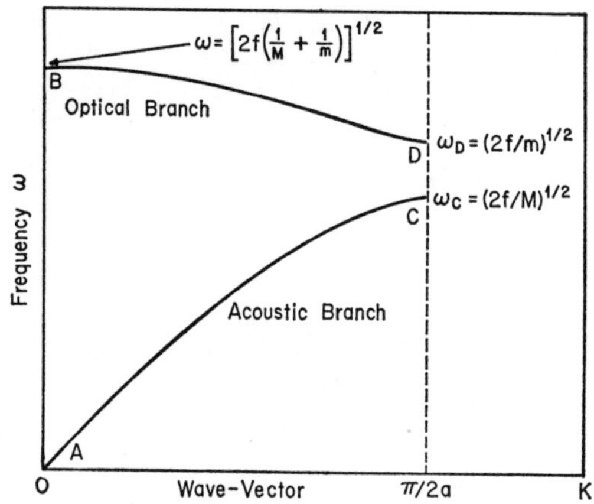

FIG. 1. The dispersion diagram for a diatomic chain.

If trial solutions of the form $u_{2j} = A\exp[i(\omega t + 2jka)]$ and $u_{2j+1} = B\exp\{i[\omega t + (2j+1)ka]\}$ are assumed, then back substitution into the equations of motion results in simultaneous equations in A and B of the form $\omega^2 MA + fB[\exp(ika) + \exp(-ika)] - 2fA = 0$ and $\omega^2 mB + fA[\exp(ika) + \exp(-ika)] - 2fB = 0$. The non-trivial solution of these equations is given by

$$\begin{vmatrix} 2f - M\omega^2 & -2f\cos ka \\ -2f\cos ka & 2f - m\omega^2 \end{vmatrix} = 0$$

and hence

$$\omega^2 = f\left(\frac{1}{M} + \frac{1}{m}\right) \pm f\left[\left(\frac{1}{M} + \frac{1}{m}\right)^2 - \frac{4\sin^2 ka}{mM}\right]^{1/2}. \quad (1)$$

The above equation is known as a *dispersion relation* and is plotted in Fig. 1.

It can be seen that the dispersion relation has two parts, usually called branches, due to the possibility of taking either positive or negative values of the second term in eqn (1). At A (Fig. 1) all the particles move in phase, while at B $\omega_{\text{max}}^2 = 2f[(1/M) + (1/m)]$ and all the m particles move one way while the M particles move the other (Fig. 2a). At C we have the highest frequency for the negative value of the square root term $\omega_{\text{C}}^2 = 2f/M$ and this corresponds to the smaller particles remaining motionless while the M particles move alternately in opposite directions (Fig. 2b). The role of the masses is reversed at D where the frequency $\omega_{\text{D}}^2 = 2f/m$ corresponds to the lowest value obtained in using the positive value of the square root term.

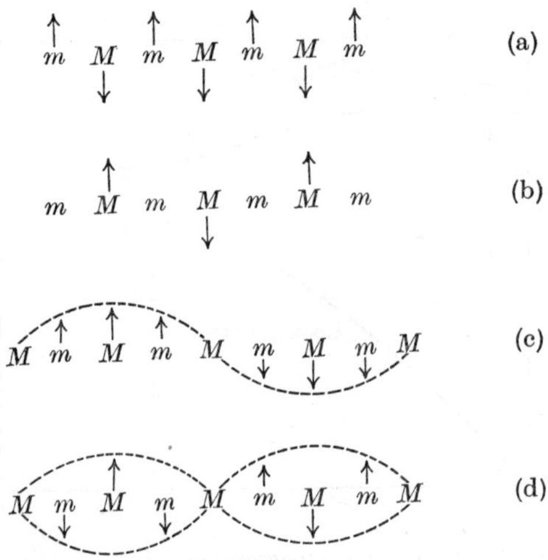

FIG. 2. Various modes of vibration of a diatomic chain. (a, b) see text, (c) a typical acoustic mode, (d) a typical optical mode.

The nature of the vibrations in the two branches differs fundamentally and it can be shown that the two types of particles are out of phase in the upper branch of Fig. 1 and in phase in the lower branch. This is depicted in Figs. 2c and 2d. In the acoustic mode (Fig. 2c) the motion is as for sound waves and hence the lower branch of Fig. 1 is termed the *acoustic branch*. In Fig. 2d the particles, if charged, produce an oscillating dipole by their movement against each other and this can interact with electromagnetic radiation. For this reason the upper branch of Fig. 1 is called the *optical branch*. The optical band width is $(2f/M)^{1/2}$. There are no solutions of eqn (1) between the acoustical and optical branches, that is between ω_{D} and ω_{C}; this region is seen to increase as $(M - m)$ increases. The lattice is said to act as a band pass filter in that it will allow certain wavelengths to be propagated but will damp out other wavelengths. So far no definition has been given for the important symbol k in eqn (1). The usual way of expressing the relationship between the frequency ω and the wavelength λ for the solutions of a wave equation in a periodic system such as a crystal lattice is by a plot of wave-vector k $(= 2\pi/\lambda)$ against frequency ω. This plot is usually nonlinear and, as mentioned before, is known as a dispersion relation.

When a periodic boundary condition is imposed the wavelength, and hence wave-vector k, is limited to a set of allowed values (the eigenvalues of the Schrödinger wave equation). The plot of $\omega - k$ repeats itself for values of k outside the range $\pm\pi/2a$, no new information being provided. The region $-\pi/2a \leqslant k \leqslant \pi/2a$ is usually referred to as the first *Brillouin zone*. The allowed values of wave-vector k, by reference to the dispersion curves, yield the permitted frequencies (or energies) of the lattice.

In extending the discussion to three dimensions, a brief mention only of the more important results is given. The more interested reader is referred to original work by Born and von Karman (1912). If the crystal is composed of N unit cells with n atoms per unit cell, then for each value of the wave-vector k there exists $(3n - 3)$ optical branches and 3 acoustical branches for each direction in the crystal. (Waves travelling in directions of high symmetry in a crystal are either transverse or longitudinal. However, in directions of no symmetry the branches may not correspond to purely longitudinal or transverse.) On each of the $3n$ branches, the frequency of a mode ($3Nn$ in number) is related to the wave-vector k, which may take any value in the Brillouin zone. This complex situation is fortunately greatly simplified in dealing with infrared radiation interacting with a crystal. The wave-vector of typical far-infrared radiation is of the order of 10^3 cm^{-1} whilst that corresponding to the Brillouin zone edge is typically 10^8 cm^{-1}. Hence, the wave-vector of the radiation may be taken as zero ($k = 0$); this is often referred to as the "long-wave limit" or "zone-centre" approximation. Consequently, the only modes of vibration active in the infrared are those in which equivalent atoms move in phase. Thus, if only $k = 0$ modes are considered then the acoustic modes have zero frequency and the spectrum consists of the $(3n - 3)$ optical modes.

B. Infrared Spectra of Ionic Crystals

In the harmonic treatment of the infrared dispersion theory of ionic crystals (Huang 1951) an expression is derived for the dielectric constant η as a function of angular frequency ω, which may be written as

$$\eta(\omega) = \eta_\infty + \frac{(\eta_0 - \eta_\infty)\,\omega_0^2}{\omega_0^2 - \omega^2}. \tag{2}$$

In this equation η_0 and η_∞ are the low- and high-frequency dielectric constants and ω_0 is the dispersion frequency. Since electromagnetic radiation has only transverse electric vector components, only transverse modes of a crystal may be excited. Thus, the dispersion frequency ω_0 can be equated to the transverse optic mode frequency ω_t. Substitution of $k = 0$ in eqn (1) yields

$$\omega_0^2 = \omega_t^2 = 2f\left(\frac{1}{M} + \frac{1}{m}\right). \tag{3}$$

It can be shown that the longitudinal frequency is given by $\omega_l^2 = \eta_0 \omega_t^2/\eta_\infty$. It should be emphasized that these equations are only valid at $k = 0$ (a non-cancelling dipole moment can only be produced by long wave ($k = 0$) phonons).

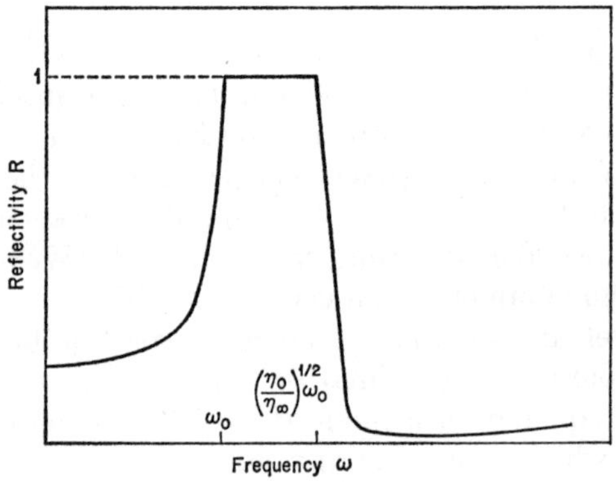

FIG. 3. The reflection spectrum of an undamped oscillator.

The harmonic picture then predicts a line absorption at the frequency of the transverse optic mode at zero wave-vector. Essentially this is borne out in practice; most diatomic ionic crystals show an intense absorption peak at a frequency given by eqn (3).

The reflectivity of a medium is given by

$$R = \frac{(n-1)^2}{(n+1)^2} \tag{4}$$

where the refractive index $n = \sqrt{\eta(\omega)}$. Fig. 3 shows a theoretical reflection spectrum of a diatomic crystal. The general features can be explained satisfactorily by considering eqns (2) and (4). As the frequency increases from zero to ω_0 the refractive index n [$=\sqrt{\eta(\omega)}$] increases from its initial value $\sqrt{\eta_0}$ up to infinity, when by eqn (4) the crystal is totally reflecting; $R = 1$. As the frequency increases further from ω_0 to $(\eta_0/\eta_\infty)^{1/2}\omega_0$ the crystal is still reflecting totally and the refractive index is imaginary. As the frequency is increased still further above the value $(\eta_0/\eta_\infty)^{1/2}\omega_0$ the refractive index is seen to decrease.

Since $\omega_t = \omega_0$ and $\omega_l = (\eta_0/\eta_\infty)^{1/2}\omega_0$ for a diatomic ionic crystal a region of total reflection is expected between ω_t and ω_l. This then explains the relative strength of the absorptions in real diatomic crystals. This selective reflection of radiation in the vicinity of the dispersion frequency is known as *Reststrahlen* behaviour and the dispersion frequency itself is known as the Reststrahlen frequency. A few Reststrahlen curves are shown in Fig. 4.

However, although all real diatomic ionic crystals in practice exhibit a Reststrahlen band (region of high reflectivity), none show 100% reflection and the majority does not follow the theoretical band contour. Eqn. (2) does not allow for absorption—an undamped oscillator

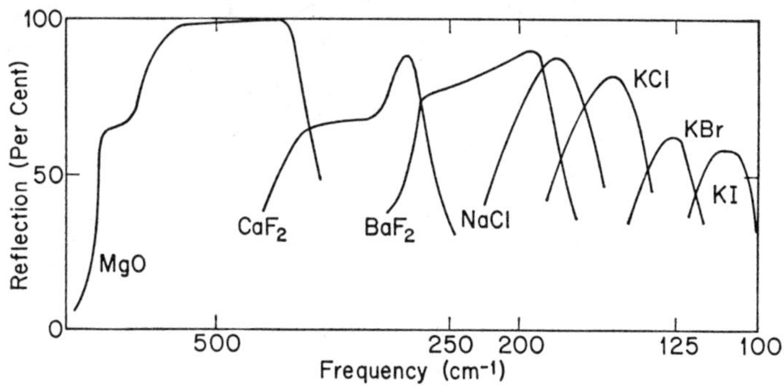

FIG. 4. The Reststrahlen behaviour (two reflections) of several ionic crystals.

cannot absorb energy since once excited to an upper state it remains there. To remedy this, a damping constant γ (or rather damping factor since it is not truly constant) is introduced, which is inversely proportional to the lifetime in the upper state. The dispersion equation then becomes

$$\eta(\omega) = \eta_\infty + \frac{(\eta_0 - \eta_\infty)}{1 - (\omega/\omega_0)^2 - i(\gamma/\omega_0)(\omega/\omega_0)} \tag{5}$$

which now covers reflection and absorption. In an absorbing medium the dielectric constant is given by $\sqrt{\eta(\omega)} = n + iK$, where n, the refractive index, and K, the extinction coefficient, are the optical constants. A more commonly encountered quantity, the absorption coefficient α, is given by $\alpha = 4\pi K/\lambda$. The reflectivity R becomes

$$\frac{(n-1)^2 + K^2}{(n+1)^2 + K^2}.$$

The effect of damping is shown in Fig. 5. This damping is an anharmonic effect and will be dealt with in a later section on anharmonicity.

Eqn (5) is found to represent the reflection spectra of several real diatomic crystals very well. A detailed examination of most spectra indicates the presence of side peaks, which are not to be expected on the harmonic model. The origin of these will be discussed in the next section, on homopolar crystals (Section C).

If the ions in an ionic crystal are treated as point charges, then it can be shown that the high-frequency refractive index of all crystals in the near infrared is greater than unity. The error is considered to result from the neglect of polarization of the ions. If the ions are assumed to

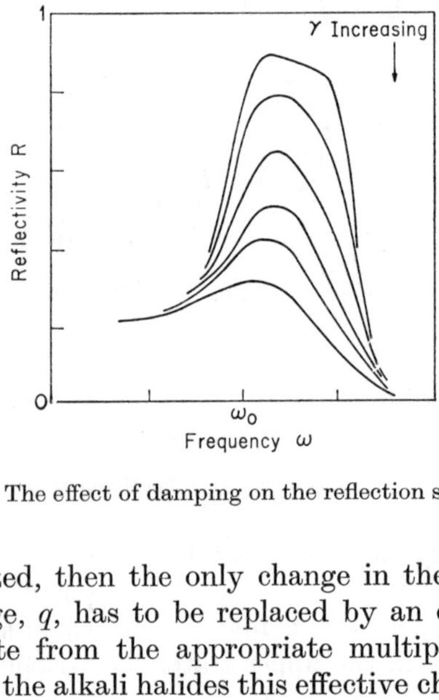

FIG. 5. The effect of damping on the reflection spectrum.

be linearly polarized, then the only change in the results is that the formal ionic charge, q, has to be replaced by an effective charge, q^*, which may deviate from the appropriate multiple of the electronic charge. In most of the alkali halides this effective charge is less than the electronic charge (see Table I). Szigeti (1950) explained this as due to localized distortions of the ionic wavefunctions in the overlap region between neighbouring ions, giving rise to a dipole moment. In the linear approximation

$$q^* = q - \partial\mu/\partial z \tag{6}$$

where μ is the distortion moment between an ion pair of separation z.

Measurements of η_0, η_∞ and ω_0 allow a value of the effective charge

$$q^* = \omega_0 \frac{(\eta_0 - \eta_\infty)^{1/2}}{4\pi} \frac{3}{(\eta_\infty + 2)} \left(\frac{V}{Ze}\right)^{1/2} \left(\frac{1}{M} + \frac{1}{m}\right)^{1/2} \tag{7}$$

to be obtained. In this equation V is the unit cell volume, Ze is the formal charge on atoms M and m.

TABLE I. Effective Ionic Charge q^* of Several Simple Ionic Crystals.

NaF	0.93
NaCl	0.74
NaBr	0.69
NaI	0.71
KF	0.88
KCl	0.80
KBr	0.76
KI	0.69
AlSb	0.53
GaP	0.58

Eqn (6) shows that, even when $q = 0$, a spectral absorption may still be observed. In covalent compounds of the type AlSb and GaP one would predict almost zero net charge per unit volume, but experimentally these compounds give rise to strong absorption with an effective charge of the order of $0.5Ze$.

C. Infrared Spectra of Homopolar Crystals

In purely homopolar crystals such as diamond, silicon, germanium, and so on, theory predicts no Reststrahlen-type absorption, since there can be no net polarization and, hence, no coupling with infrared radiation. (Homopolar crystals do, on the other hand, exhibit a Raman spectrum.) The above compounds, however, do possess infrared spectra, but unlike the Reststrahlen-type spectrum, these are relatively weak with much more structure. The origin of such spectra, and of the side peaks observed with ionic crystals, is due to anharmonic effects and will be discussed in the section on anharmonicity (Section E).

D. Infrared Spectra of Molecular Crystals

If a crystal contains only atoms, then the only modes possible are lattice modes. However, if a crystal contains molecular groupings, the forces within a group being stronger than the forces between groups, then it is possible to distinguish two types of vibration. Modes arising from the movement of atoms within a group are called internal modes while modes in which one group moves relative to another group are known as external or lattice modes. Due both to the weaker intermolecular forces and the heavier masses involved, the external modes will be expected at lower frequencies than the internal modes. In a crystal, the degrees of freedom allotted to the rotation of an isolated molecule are expended in modes of vibration about an equilibrium orientation, while the translation of a free molecule is transferred in a crystal into modes of vibration about an equilibrium position. These forms of vibratory motion in a crystal are known as rotatory (or librational) lattice modes and translational lattice modes respectively. In general, the rotatory modes will occur at higher frequencies and with greater intensity than the translatory modes. The situation in a crystal with p molecules per unit cell and q atoms per molecule is thus:

		Approximate frequency range
Internal molecular modes	$p(3q-6)$	200–4000 cm^{-1}
Rotatory lattice modes	$3p$	100–300 cm^{-1}
Translatory lattice modes	$3p-3$	below 100 cm^{-1}
Total =	$(3pq-3)$	

The normal modes of a molecular crystal will now be considered in detail. The potential energy of a molecule in the vapour phase can be expressed solely in terms of internal parameters describing displacements from the equilibrium configuration, provided that it is assumed that intermolecular interactions in the vapour phase are so small as to be negligible. However, in the solid phase (and to a lesser extent in the liquid phase), the interactions between molecules produce additional terms in the expression for the potential energy. In order to interpret solid-state spectra, the effects of these additional terms must be considered.

The harmonic approximation implies that, (a) the potential energy contains only quadratic terms, and (b) only linear terms need be included in the expansion of the dipole moment as a function of the normal coordinates, that is $\mu = \mu^0 + (\partial\mu/\partial Q)Q$. In this approximation it can be shown that the infrared spectrum of a crystal should consist of relatively few lines. All degrees of freedom were included in the theory so that none (such as rotational degrees of freedom in the vapour state) remain to yield a band, rather than a line, by coupling with the vibration. These fundamentals should arise only from those motions in which equivalent atoms in each unit cell move in phase.

The potential energy in the harmonic oscillator approximation may be written as

$$V = \sum_j (V_j^0 + V_j') + \sum_j \sum_i V_{ij}'' + V_L + V_{Lj}'''$$

where V_j^0 represents the potential energy of the jth free molecule. The associated force constants determine the fundamental frequencies and the molecular symmetry determines the spectral activity of the various modes. V'_j represents the change in V^0_j due to the equilibrium field of the crystal. V''_{ij} represents the interaction between the displacement coordinates of the ith and jth molecules (cross-terms between internal coordinates of different molecules). V_L represents the lattice potential and includes the terms involving relative changes in the centres of gravity and orientations of the molecules; V'''_{Lj} represents the interaction between internal and lattice coordinates.

The term V'_j causes the potential energy of the molecule in the crystal V_j $(= V^0_j + V'_j)$ to differ from that in the vapour phase V^0_j and the resulting effects are called *static field* effects. These are a direct measure of the influence which the crystal lattice, in its equilibrium configuration, exerts upon a molecule. They can produce shifts in frequency and may lead to a splitting of a degenerate vibration since the term V_j has the symmetry of the jth molecule on its site in the crystal lattice (the so-called *site symmetry*) and this is usually lower than that of the free molecule in the vapour. It is even possible that this relaxation of the selection rules will cause a fundamental, forbidden in the free molecule, to become active in the site group analysis.

Example 1.

The various BCl_3 frequencies (cm^{-1}) given below indicate typical static field shifts ($\Delta\nu$).

	Vapour	Crystal	Solid soln.	$\Delta\nu$ (vapour–crystal)
ν_2 (B^{11})	460	447	445	13
ν_2 (B^{10})	480	469	466	11
ν_3 (B^{11})	956	926	936	30
ν_3 (B^{10})	995	962	980	33

Example 2.

In cyclopropane the symmetry of the free molecule is D_{3h} while in the crystal the site symmetry is C_s. A CH_2 deformation mode of species a'_1 is observed at 1454 cm^{-1} in the Raman spectrum of the vapour but the same mode is forbidden in the infrared spectrum. However, the infrared spectrum of the solid shows a band at 1454 cm^{-1}. Another CH_2 deformation mode, but of the e' symmetry species, is observed at 1442 cm^{-1} in the infrared spectrum of the vapour. The corresponding solid state spectrum shows a doublet at 1424–1434 cm^{-1}. Both of the above effects are a consequence of the lowering of the symmetry in going from the vapour to the crystal

The second term in the above expression for the potential energy, that is $\sum_j \sum_i V''_{ij}$, produces so-called *dynamic field* effects. The term couples the vibrations of all the molecules, but if this coupling is small then first-order perturbation theory may be applied to the problem, with the result that only coupling between equivalent modes of the various molecules need be considered. If there are p molecules per unit cell (the smallest volume of the lattice from which the entire crystal may be generated by translations alone), the result is that each non-degenerate mode of an individual molecule may be split into a maximum of p components. This splitting is known as *correlation field* splitting. It can be shown that the weighted mean value of all the components should be the frequency at which the vibration would occur if only the static field operated. This last statement is not strictly true for the following reason. The terms $\sum_j \sum_i V''_{ij}$ may be separated into two types: (a) those which involve molecules which are translationally non-equivalent, giving rise to correlation field splittings, and (b) those terms which couple translationally equivalent molecules and produce a frequency shift which is independent of the correlation field splitting. This separation of the $\sum_j \sum_i V''_{ij}$ terms is based upon the criterion for infrared activity, that is infrared active modes in a crystal are such that corresponding movements in all unit cells are in phase. Thus, the mean value of the correlation field components occurs at a frequency ν_X which differs from the gas-phase frequency ν_G by an amount $\Delta\nu_T = \nu_G - \nu_X = \Delta\nu_m + \Delta\nu_c$. The first part of the frequency shift, $\Delta\nu_m$, arises from the potential energy term V'_j and has already been discussed. The second part, $\Delta\nu_c$, arises from the $\sum_j \sum_i V''_{ij}$ terms which couple translationally equivalent molecules.

Example 3.

With reference to Example 1 it can be seen that for BCl_3:

	$\Delta\nu$ total (cm^{-1})	$\Delta\nu_m$ (cm^{-1})	$\Delta\nu_c$ (cm^{-1})
(B^{11})	13	15	−2
(B^{10})	11	14	−3
(B^{11})	30	20	10
(B^{10})	33	15	18

Example 4.

For the HBr molecule, the following frequencies (cm^{-1}) have been observed:

Vapour	Crystal	HBr/DBr	HBr/HCl
2560	2438 2404	2422	2440

The correlation field splitting is thus 34 cm^{-1} and $\Delta\nu$ total is 139 cm^{-1}.
For the case of HBr/DBr: $\Delta\nu_m = 138$ cm^{-1} and $\Delta\nu_c = 1$ cm^{-1}
For the case of HBr/HCl: $\Delta\nu_m = 120$ cm^{-1} and $\Delta\nu_c = 19$ cm^{-1}.

By including a term in the potential energy which couples together vibrations of different molecules it is necessary to analyse the spectrum according to the symmetry of the unit cell, rather than the symmetry of a molecular site. It can be shown that the number of frequencies active in the infrared (or Raman) calculated from the local (site) symmetry, including each non-equivalent type of molecule once, is equal to the number calculated on the unit cell basis, which is also equal to the number allowed for the entire crystal. Polarization data can only be calculated on the unit cell basis.

Three commonly occurring types of symmetry which are of importance are the molecular symmetry (symmetry of the free molecule as given by its point group), the site or local symmetry, and the unit cell or factor group symmetry. The site group must be a subgroup of both the factor group and the molecular point group, and consequently the following generalized "rules" exist: (1) every vibration active under the site group gives rise to at least one active component in the crystal, and (2) any vibration degenerate under the site group remains so in the crystal. This means that the correlation field effect does not split a degeneracy which survives the site group. A further but obvious "rule" is, (3) that any vibration inactive under the site group remains inactive in the crystal. Fig. 6 depicts diagrammatically a hypothetical situation to illustrate the above "rules".

Now (1) if a is active, then at least one of a' is active. If a is inactive, then both of a' are inactive. Because of rule (2) we expect e' to consist of two lines and not four. The weighted mean frequency of $a'(e')$ lies at the frequency of $a(e)$.

The effect of the potential energy term V'''_{Lj} on the internal vibrations may, in most cases, be considered negligible, due to the large frequency separation of the lattice and internal frequencies. When the frequency of an internal mode is comparable to the lattice frequency (usually less than 200 cm^{-1}) then the approximation would no longer be valid. However, unless the above situation occurs, or the coupling is excessively large, a satisfactory approximation to the internal frequencies may be obtained by omitting the V'''_{Lj} term. The effects of this term will be discussed in Section E.

If, in addition to neglecting the V'''_{Lj} term, we also neglect the term $\sum_j \sum_i V''_{ij}$, so that the potential energy takes the form $V = \sum_j (V_j^0 + V'_j) + V_L$, then the vibrational problem is separable into a part associated with the lattice vibrations and another part dealing with the vibrations of the molecules. The vibrational frequencies of the molecules would be the same as for an isolated molecule except for a shift due to the term V'_j. The *oriented gas* model, in which the molecules are considered to be

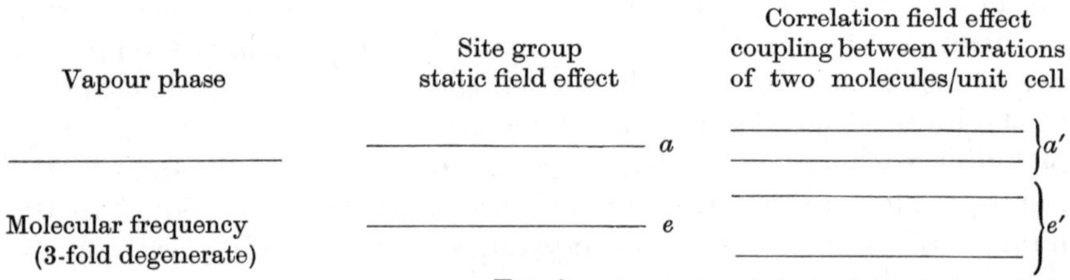

FIG. 6.

oriented in fixed positions with zero intermolecular forces operative, is tantamount to the above approximation if V'_j is small.

E. Anharmonic Effects

The effects of anharmonicity will be separated into, (1) those occurring in ionic and homopolar crystals, and (2) those which are manifested in molecular crystals.

1. *Anharmonicity Effects in Ionic and Homopolar Crystals*

In the general theory of small vibrations it can be shown that if the kinetic and potential energies of the system are expressed in terms of a certain set of coordinates, then the system behaves as a series of uncoupled oscillators. Using these *normal coordinates* the vibrations of a crystal can be expressed in terms of simple harmonic oscillators which are independent, since no cross terms linking different coordinates occur. This is true only in the harmonic approximation which, therefore, predicts the vibrational spectrum of a crystal to consist of a series of sharp, temperature-independent absorptions at the $k = 0$ transverse optic frequencies.

In practice, however, while the number of such absorptions is correctly predicted, they are significantly broadened (damped), and slightly temperature-dependent. In addition, weak features appear in the spectrum of ionic crystals and, in homopolar crystals, many weak absorptions are found, totally in disagreement with the harmonic picture.

The above effects can be explained as due to combination bands which appear as a consequence of anharmonicity. The mechanism by which two or more phonons combine to interact with the radiation has been attributed to the presence of cubic terms in the potential energy. The interaction occurs through the dipole moment of the fundamental mode. To account for the absorptions observed in homopolar crystals (where there is no dipole moment), Lax and Burstein (1955) postulated that two phonons could interact directly with the radiation *via* second-order electric moment terms. One vibrational mode is thought to induce charges on the atoms and a second mode simultaneously vibrates these charges, producing an electric moment whose magnitude will be the sum or difference frequency of the two interacting phonons.

The broadening of the main absorption bands in ionic crystals was explained as due to the presence of anharmonic terms (cubic, and so on) in the potential energy associated with the lattice vibrations (Blackman 1933). Lax and Burstein have shown that the presence of side peaks in spectra of ionic crystals is probably due both to the above mechanism and to the combination bands produced by the presence of second and higher-order terms in the electric moment expansion in terms of ionic displacements.

In homopolar crystals, the absorption must arise as a consequence of the second-order electric moment terms. (No electric moment can be produced unless a charge deformation is produced, presumably by modes of short wavelength, that is zone boundary phonons.) The presence of anharmonic terms in the potential energy will cause further broadening of the absorption. Absorptions in heteropolar covalent crystals may be similarly explained.

It is not yet clear which of the two mechanisms (third-order potential energy terms or second-order dipole moment terms) is more important in the case of ionic crystals, since both predict essentially the same shape and temperature dependence for the absorption, which is determined mainly by the frequency distribution of the lattice vibrations. Kleinman (1960) concluded that, at least in gallium phosphide, "it is not necessary to invoke the second-order electric moment mechanism to explain the observed infrared absorption bands".

The combination bands actually constitute a continuous absorption, since essentially there are no selection rules but "singularities" in the phonon frequency distribution produce peaks in the absorption. These singularities occur wherever the dispersion curve is such that the gradient of ω with respect to k is zero. Such so-called *critical points* are expected to occur at or near the edge of the Brillouin zone. To interpret the spectra of crystals, therefore, in the anharmonic approximation, it is necessary to consider all the branches on the dispersion diagram, at all values of the wave-vector k from zero to the zone boundary. In the harmonic model, only the transverse optic branch is considered at $k = 0$. The typical behaviour of optical branches (shown in Fig. 7) is that at low k values and k values near the zone boundary, the frequency is only

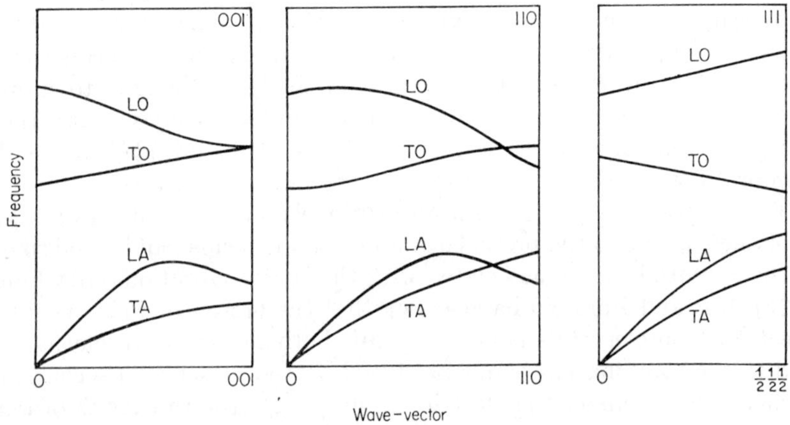

FIG. 7. Typical lattice dispersion curves. The curves above are those calculated for NaI using a shell model (Woods *et al.* 1960). In each direction the transverse branches are doubly degenerate and the longitudinal branches singly degenerate.

slightly wave-vector dependent, while at intermediate k values the frequency variation with k is more pronounced.

One of the most important concepts encountered in a discussion of the spectra of anharmonic crystals is that of the "density of states"—the number of lattice modes within a small frequency range, centred on the frequency of interest. How this quantity varies with the frequency is of prime importance in determining the features of a spectrum. Any branch on a dispersion diagram will contribute strongly to a density-of-states curve at any frequency at which the gradient of ω with respect to wave-vector is almost zero. Bearing in mind that the above, and indeed any, dispersion curve refers to waves travelling in a certain single direction in a crystal, extension of this idea to three dimensions (or k-space) leads to the conclusion that whenever the quantity grad $[\omega(k)]$ is almost zero for any branch then the density-of-states curve should exhibit a characteristic feature (critical point). It can be shown that almost all of these critical points can be located simply by applying symmetry arguments to the crystal structure. Most are found either at $k = 0$ or at specific points on the Brillouin zone surface in high symmetry directions. Figure 8 shows the Brillouin zone and major critical points for the diamond-type structure.

As pointed out earlier the spectra in the anharmonic approximation are "multiphonon" spectra in so much as an incident photon excites two or more phonons simultaneously. The conservation of both energy and momentum must be observed so that if the photon (phonon 1, phonon 2) has energy $\hbar\omega(\hbar\omega_1, \hbar\omega_2)$ and momentum $\hbar k(\hbar k_1, \hbar k_2)$ then $\hbar\omega = \hbar\omega_1 + \hbar\omega_2$ and $\hbar k_1 + \hbar k_2 = \hbar k$. For infrared radiation $k = 0$, so that $k_1 = -k_2$. Thus,

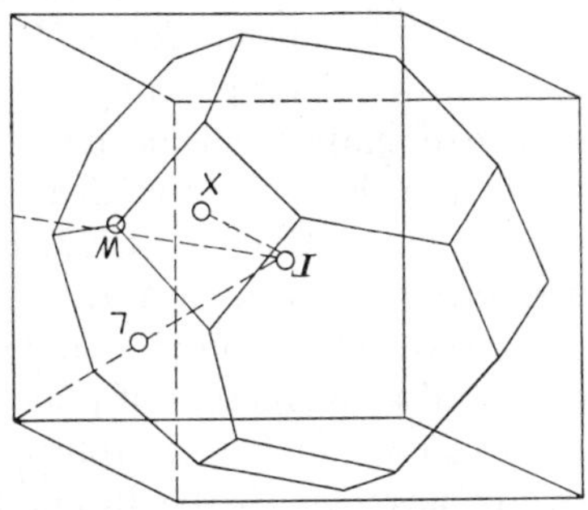

FIG. 8. The Brillouin zone for the diamond (and rock salt) structures, showing critical points at Γ, X, L and W. $\Gamma = 0, 0, 0$, $\mathrm{X} = 1, 0, 0$, $\mathrm{L} = \frac{1}{2}, \frac{1}{2}, \frac{1}{2}$, $\mathrm{W} = 1, \frac{1}{2}, 0$.

addition of the branches of the one-phonon curves, in pairs, taking corresponding frequencies for each k value within the zone, will result in a "combined" dispersion curve which represents the spectrum's two-phonon contribution. Corresponding to a series of these combined dispersion curves would be a combined density-of-states curve for two-phonon excitation, which would usually show critical point discontinuities.

For diamond-type structures, a particularly flat combined dispersion branch (and hence, high density of states) is obtained from summing the transverse optic and transverse acoustic single phonon branches. In practice, a characteristic peak is found in the spectra of diamond, silicon and germanium at the appropriate frequency.

As will be shown later, the form of the absorption spectrum is dependent on the density of states $S(\omega)$ and on an amplitude factor [designated $M^2(\omega)$] which usually is a smoothly varying function of frequency ω. Thus, it is expected that all the features in the density of states, including all slope discontinuities associated with critical points, should be the only sharp kinks appearing in the spectrum. Hence, although the spectral maxima may not occur at the same frequency as the density-of-states maxima, these kinks should appear at exactly the "correct" frequency.

This makes it possible to analyse spectra to obtain values of phonon frequencies at specific points in the Brillouin zone, rather than just those for density maxima, as is more usual. Key points in the dispersion curves are thus located.

The simplest method of analysing two-phonon spectra is to assume four characteristic phonons, designated TA, LA, TO, LO, which correspond to the zone boundary values of the transverse and longitudinal, acoustic and optical branches, respectively. The frequencies observed experimentally are then fitted to combinations of these four phonon energies. Removal of the two-fold degeneracy of the transverse branches is sometimes postulated in order to empirically fit the data.

An extension of this type of analysis has been made by a consideration of the effects on the spectrum due to the critical points—the so-called "critical point analysis" method. Cowley (1962) has given the theory dealing with the lattice dynamics of a crystal with polarizable atoms. The equations, which when solved provide the $3p$ ($p =$ number of atoms per unit cell) eigenvalues and eigenvectors of phonons of a given wave-vector, involve force-constant matrices coupling different unit cells. The long range forces are assumed to be interactions between point monopoles and dipoles and are readily evaluated. However, at short ranges the above assumption is invalid and the forces are introduced specifically. The values of the short range force constants (assumed to be independent of wave-vector) utilized in the calculation are selected so as to fit experimental data such as dielectric and elastic constants.

It is found that when the short range forces consist of nearest neighbour interactions only, the calculated dispersion curves are normally within 15% of the experimentally determined curves. Dolling (1962) has shown that substantial improvement can result by including second neighbour forces.

Several multiphonon analyses of some detail have been performed by Woods *et al.* (1960), Cochran (1959), and Johnson and Cochran (1962), using the results of theoretical studies of the selection rules operative in such processes (Birman 1963; Loudon 1964).

2. *The Effects of Anharmonicity in Molecular Crystals*

The inclusion of anharmonic potential energy terms has the effect of introducing the possibility of overtones and combination frequencies into the spectrum, which in the harmonic approximation should consist of a limited number of narrow fundamentals, whose intensities are temperature-independent.

Concerning combinations between internal vibrations of molecules in crystals, the situation is analogous to that of isolated molecules in the vapour (case 1: internal + internal). The most important qualitative difference between the case of a crystal and of a free molecule is that combination bands between lattice modes and internal vibrations of a molecule are allowed (the effect of the V'''_{Lj} term). Due to the fact that most lattice modes occur at low frequencies (less than 200 cm^{-1}) such combination bands tend to broaden the harmonic narrow line absorption into a band, rather than appear as separate absorptions. A forbidden fundamental may be activated by its combination with a lattice mode (lattice and internal modes may only combine if they have the same symmetry) whence absorption may occur in the immediate vicinity of the forbidden fundamental, in violation of the harmonic selection rules (case 2: internal + lattice). The third type of combination which may occur in a molecular crystal is the one between lattice modes themselves. This will not be discussed further here since essentially it is analogous with the multiphonon absorptions treated earlier (case 3: lattice + lattice).

3. *Defect-Induced One-Phonon Absorption*

If defects are produced in a crystal lattice either by replacing some of the atoms by others of different mass and force constant or by introducing extra charge on some of the atoms, then the symmetry of the lattice is reduced and the motions of the normal modes of the lattice are modified, especially near the actual defects. Some of these modified modes (which when unmodified in the pure crystal do not interact with infrared radiation) produce extra linear terms in the dipole moment expression and thus cause additional one-phonon absorption. Because of this, the so-called "defect-induced" spectra provide more information concerning the pure crystal than do the "pure crystal" spectra themselves, providing that the defect density is not excessive.

The theory of the effects of changes of mass and force constant on the normal modes of defective crystals has been treated by several authors (Wallis and Maradudin 1960; Szigeti 1963; Dawber and Elliott 1963a, b) and only the more relevant consequences are dealt with below.

If the atom replaced in a monatomic crystal is lighter than the defect atom, then all the crystal modes are found to remain "non-localized", that is the motion is propagated by the lattice. However, when the atom replaced is heavier than the defect atom, then a few of the modes are "localized" around the defect, the remaining modes being non-localized. Obviously, the movement of atoms in non-localized modes in a defective crystal is very similar, away from the defect centre, to the normal modes in the pure crystal. The motion is strongly perturbed, however, in the immediate neighbourhood of the defect. This perturbation is a function of the defect mass and governs the intensity with which the absorption appears. The frequency at which this absorption occurs can be shown to obey the expression $\Delta\nu = \nu_{\text{pure}} - \nu_{\text{defect}} \propto m_{\text{defect}}/m_{\text{crystal}}$; hence, these non-localized modes of the defect crystal will appear in the same ranges as would the normal modes of the pure crystal and, furthermore, will be characteristic of the impurity concerned. This strong dependence of the absorption on the nature of the defect illustrates the importance of studying systems in which the defect is simple and known. It can now be appreciated why the study of defect crystals is proving so effective in obtaining information concerning the frequency spectrum of pure crystals.

The localized modes, which give rise to sharp lines, appear in the spectrum at frequencies outside the above mentioned allowed ranges. Usually they lie above the ranges of the non-localized modes.

In a diatomic lattice it is possible to obtain localized mode frequencies between the acoustic and optical bands. It may be shown that if the lighter of the two kinds of atoms is replaced by a heavier defect atom, then only the localized frequencies between the acoustic and optical bands are observed. If the heavier of the two kinds of atoms is replaced by a lighter defect atom, then in addition to those found between the acoustic and optical bands, localized modes are found above the optical branches.

It can also be shown that for a crystal containing a sufficiently small number of randomly distributed defects, the absorption coefficient $\alpha(\omega)$, associated with the non-localized or band modes, is given by:

$$\alpha(\omega)_{\text{BM}} = \frac{2\pi^2 D}{3nc} S(\omega)\, M^2(\omega) \tag{8}$$

where D is the density of defects, n is the refractive index of the host crystal, $M(\omega)$ is the dipole moment of the non-localized mode of frequency ω, and $S(\omega)$ is the density of states at this frequency. Hence, the non-localized mode spectrum will reflect the density of states $S(\omega)$, modulated by a frequency-dependent dipole moment term.

An interesting phenomenon occurs for the case of defect atoms heavier than the host atoms. In this case the term $M^2(\omega)$ may peak within the range of band modes and produce quite a sharp peak in the spectrum. These peaks are referred to as "in-band resonances" or "approximate local modes" by Brout and Visscher (1962) and obviously depend strongly on the nature of the defect. The more interested reader in the theoretical aspects of crystal defect spectra is referred to the original papers by Wallis and Maradudin (1960), Szigeti (1963) and Dawber and Elliott (1963a, b).

F. Summary

The study of Reststrahlen-type bands provides direct information concerning the optically active lattice vibrations (or phonons) of zero wave-vector ($k = 0$) only. This type of single-phonon absorption, (a) only occurs with ionic crystals, (b) is a consequence of the conservation of momentum in the photon–phonon interaction, and (c) cannot provide much information concerning the phonon spectrum due to the restriction on wave-vector.

Multiple phonon absorptions may occur in both ionic and homopolar crystals. The interaction of a photon with two or more phonons allows phonons of any wave-vector to contribute since the conservation of momentum can be satisfied among the phonons. Phonon frequencies at various points on the dispersion curves may thus be obtained.

Defect-induced single-phonon absorption is due to the presence of an impurity in a crystal which, when pure, exhibits no absorption. In essence, the impurity absorbs the phonon's momentum and so facilitates single photon–phonon interactions for phonons of any wave-vector.

These two latter techniques are now being extensively used to study phonon spectra in conjunction with the more direct method of neutron spectroscopy.

II. Applications

A. Determination of Optical Constants

The optical properties of a medium are usually defined by its so-called "optical constants": n = index of refraction and K = index of absorption. Neither of these quantities, however, is constant, but varies with frequency. The quantity $N = n + iK$ is known as the "complex" index of refraction.

If the dispersion formula

$$\eta(\omega) = \eta_\infty + \frac{(\eta_0 - \eta_\infty)}{1 - (\omega/\omega_0)^2 - i(\gamma/\omega_0)(\omega/\omega_0)} \tag{9}$$

is expanded into its real and imaginary parts, we have

$$\eta'(\omega) = n^2 - K^2 = \eta_\infty + \frac{(\eta_0 - \eta_\infty)[1 - (\omega/\omega_0)^2]}{[1 - (\omega/\omega_0)^2]^2 + (\gamma/\omega_0)^2(\omega/\omega_0)^2} \tag{10}$$

and

$$\eta''(\omega) = 2nK = \frac{(\eta_0 - \eta_\infty)(\gamma/\omega_0)(\omega/\omega_0)}{[1-(\omega/\omega_0)^2]^2 + (\gamma/\omega_0)^2(\omega/\omega_0)^2} \tag{11}$$

where ω_0 = dispersion frequency and γ = damping factor.

Optical investigations of solids usually take the form of measuring either the transmission spectrum or the reflection spectrum, at normal incidence, over a wide range of frequency, and then evaluating the optical constants from the above dispersion formulae.

The reflectance, R, of an absorbing medium is given by

$$R = \frac{(n-1)^2 + K^2}{(n+1)^2 + K^2} \tag{12}$$

while the transmission, T, of a sample whose thickness, d, is large compared with the wavelength, is given by

$$T = \frac{(1-R_1)^2 \exp(-\bar{K}d)}{1 - R_1^2 \exp(-2\bar{K}d)} \tag{13}$$

where $\bar{K}$ is the absorption constant and is given by $\bar{K} = 4\pi K/\lambda$, and R_1 is the reflectance between vacuum and a medium characterized by optical constants n and K. The reflectance and transmission are related by the equation $R = R_1(1 + T\exp(-\bar{K}d))$.

If the index of absorption is small, the transmission is measured by K evaluated from eqn (13), using $R_1 = (n-1)^2/(n+1)^2$.

With a large index of absorption, transmission measurements become difficult to obtain and, in practice, the reflectance is studied.

If the values of the low- and high-frequency dielectric constants η_0 and η_∞ are known then, for a series of (γ/ω_0) values, eqns (10), (11) and (12) may be used to evaluate R as a function of (γ/ω_0). A plot of $1/R_{max}$ against (γ/ω_0) gives an approximately linear curve (Haas and Ketelaar 1956), from which may be read the value of (γ/ω_0) corresponding to the measured R_{max}. Values of the optical constants n and K then follow from substitution into eqns (10) and (11).

The reflectance of ZnS is shown in Fig. 9. The associated optical constant curves derived from this are shown in Fig. 10.

The maximum of K occurs very near ω_0 but it can be shown that the quantity $nK\omega$ is maximized at $\omega = \omega_0$ and thus the most useful plot to determine ω_0 is one of η'' against ω (Fig. 11).

It has been shown that for small damping factors (γ/ω_0), ω_0 can be determined from the relation

$$\frac{\omega_{max}}{\omega_0} = \left[1 + \frac{\eta_0 - \eta_\infty}{6\eta_0 - 4}\right]^{1/2}.$$

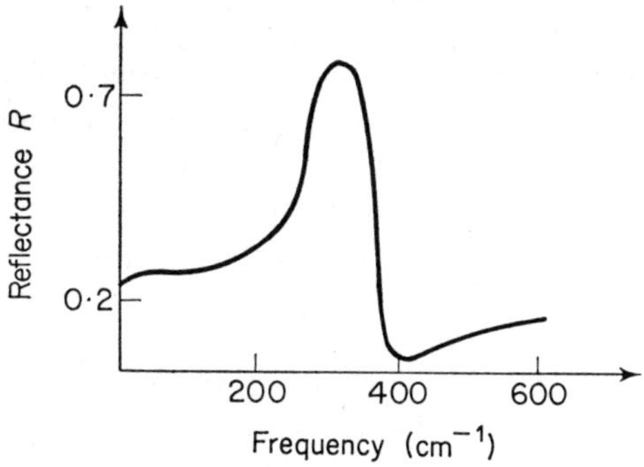

FIG. 9. The reflectance of zinc sulphide.

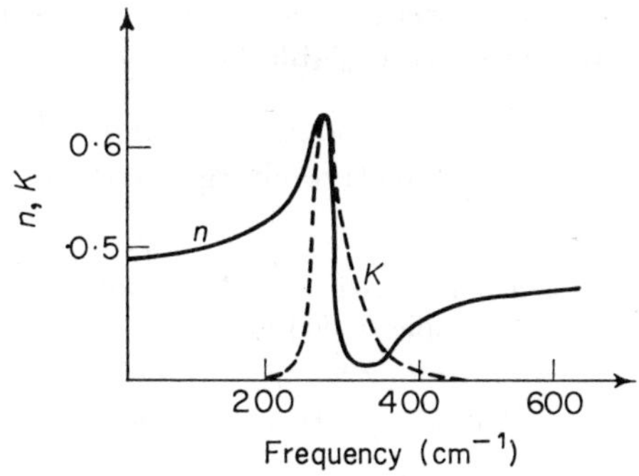

FIG. 10. The optical constants of zinc sulphide.

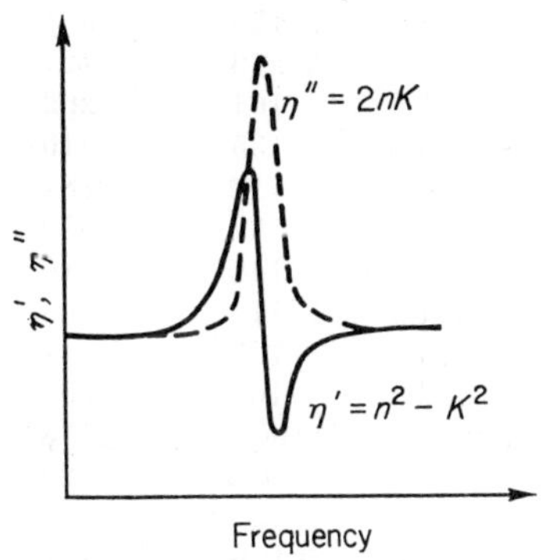

FIG. 11. The variation of η' and η'' with frequency for zinc sulphide.

As has been mentioned earlier, ω_0 is the zone centre transverse optical mode frequency, ω_t. The longitudinal optical mode frequency ω_l can be found from the equation due to Lyddane, Sachs and Teller (1941), that is $\omega_l = (\eta_0/\eta_\infty)^{1/2}\,\omega_t$.

B. Simple Ionic Crystals

The alkali halide crystals have been the subject of numerous optical investigations and in each case the appearance of a single strong absorption or reflection band in the far-infrared region has essentially borne out the harmonic model. With the introduction of a frequency-independent damping factor, the observed transmission or reflection spectra close to the band can be adequately described by the usual Lorentz type dispersion formula. The associated parameters η_0, η_∞, ω_0 and γ for several alkali halides together with other simple ionic crystals are shown in Table II; also given are the values of $\omega_l = (\eta_0/\eta_\infty)^{1/2}\omega_0$ and q^*/Ze (calculated from eqn 7).

Data for Table II is from Tables III, IV, and V of the paper by Martin (1965). The ω_0 values in parentheses refer to measurements at 4°K. The following may be noted from Table II:

(a) the γ values are all low, corresponding to high reflectance;
(b) the values of q^*/Ze, the effective charge, may be some 30% below

TABLE II. The Parameters Characterizing the Main Lattice Resonances of Some Simple Ionic Crystals.

Compound	η_0	η_∞	ω_0	ω_l	γ	q^*/Ze
LiF	8.96	1.90	304	660	0.053	0.81
LiCl	11.95	2.75	204	425		
LiBr	13.25	3.16	173	354		
NaF	5.10	1.72	246	424	0.065	0.83
NaCl	5.91	2.31	164	262	0.055	0.76
NaBr	6.38	2.63	135	210	0.067	0.75
NaI	7.26	3.03	117	181	0.18	0.75
KF	5.46	1.85	192	330		
KCl	4.87	2.16	144	216	0.064	0.82
KBr	4.91	2.34	116	168	0.034	0.79
KI	5.10	2.59	103	144	0.048	0.75
RbF	6.48	1.93	160	293		
RbCl	4.92	2.19	119(127)	178	0.076	0.84
RbBr	4.87	2.33	90(97)	130	0.056	0.82
RbI	4.94	2.62	75(82)	103	0.058	0.75
TlCl	32.3	5.1	64(66)	161	0.25	0.81
TlBr	30.6	5.41	48(48)	114	0.17	0.84
AgCl	11.1	4.04	103(122)	171	0.40	0.66
AgBr	13.1	4.62	81(93)	136	0.30	0.70
CaF_2	6.7	2.05	257	464	0.02	
SrF_2	6.6	2.07	217	387	0.02	
BaF_2	7.2	2.16	184	336	0.02	

the formal electronic charge. This is probably due to a redistribution of charge during motion, rather than an actual loss of charge;
(c) the dispersion frequency ω_0 would be expected to increase as the temperature decreased, due to contraction of the lattice and consequent

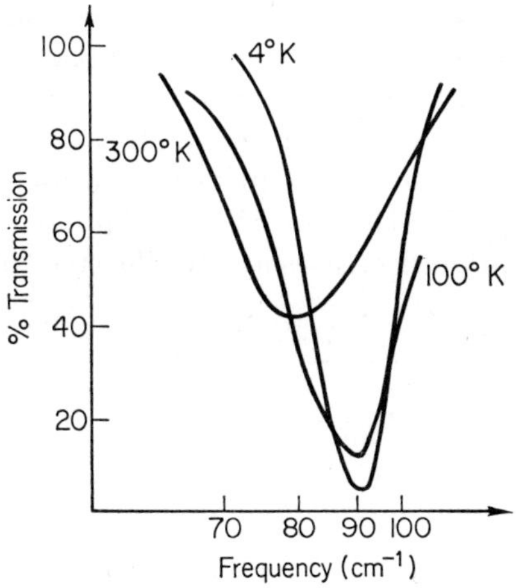

FIG. 12. The effect of temperature on the dispersion frequency of AgBr.

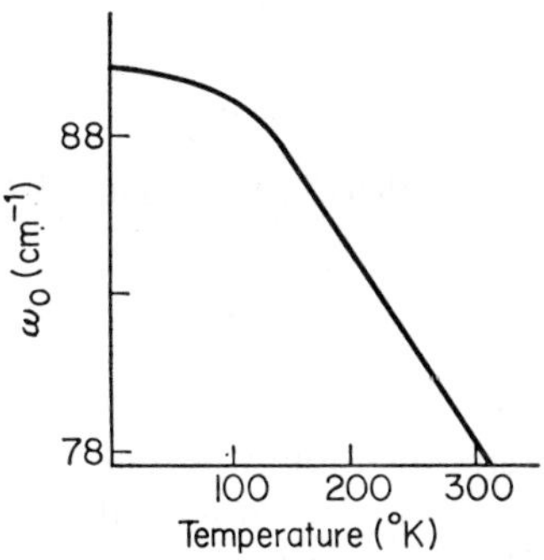

FIG. 13. The variation of ω_0 with temperature for AgBr.

increase in lattice force constants. This is in fact observed for the few examples quoted in Table II, where temperature effects were studied. The variation of ω_0 as a function of temperature is shown for AgBr in Figs. 12 and 13 (Jones *et al.* 1961).

Cowley (1963), however, has shown that two factors, with opposite sign, have to be accounted for in dealing with the variation of ω_0 with temperature. It appears that for TlBr these two factors cancel,

Fig. 12 also shows that the band width is decreasing as the temperature is lowered.

Hass (1960) examined NaCl in the temperature range 300–985°K and found that, consistent with theory, γ varied as T^2. Fig. 14 shows the variation of γ with temperature for AgBr (Jones *et al.* 1961).

The minor absorptions which appear on the high-frequency side of the main absorption in the spectra of most alkali halide type crystals (not quoted in Table II) are attributable to multiphonon processes. The extent of such absorption is less in the alkali halides than in compounds of the group II–VI type, which in turn exhibit less structure than compounds of the type III–V or IV–IV. Apparently, the amount of multiphonon absorption decreases as the effective charge increases.

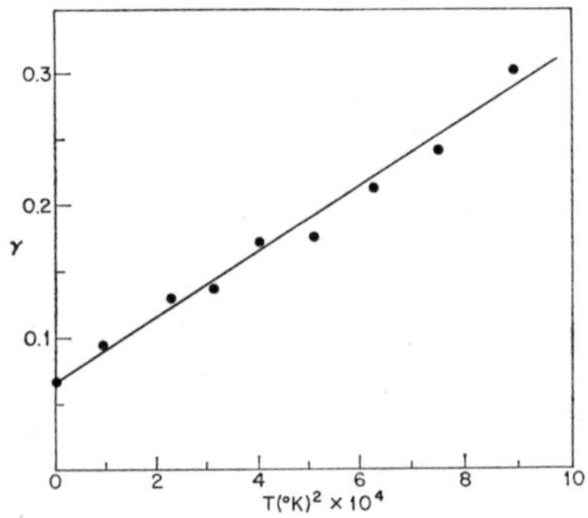

FIG. 14. The variation of γ with temperature for AgBr.

C. Multiphonon Absorptions

1. *Group II–VI Type Compounds*

Although the transverse and longitudinal optical phonon frequencies at the zone centre have been determined for several type II–VI compounds (see Table III), the small number of studies of multiphonon absorption in such compounds reflects the limited amount of structure usually observed. The data, where available, have only been analysed on a simple four- or six-phonon scheme.

For ZnSe (zinc-blende structure) eight combination bands were assigned by Mitra (1963) using a four-phonon scheme. In CdS (hexagonal cubic) a six-phonon scheme was necessary (Marshall and Mitra 1964). Zinc sulphide can occur in two different forms—a hexagonal and a zinc-blende modification—which give rise to slightly different spectra. The multiphonon absorptions in the hexagonal type have been analysed on a six-phonon scheme while for the zinc-blende modification two possible assignments have been given by Marshall and Mitra (1964).

Table III summarizes the zone centre and zone boundary phonon frequencies of some group II–VI type compounds.

TABLE III. Zone Centre and Zone Boundary Phonon Frequencies (cm^{-1}) of Some II–VI Type Compounds.

Compound	Structure	ω_t	ω_l	q^*/Ze	LO	TO	LA	TA
ZnSe	Z – B	209	250	0.71	208	212	162	87
CdS	H E X	241	319	1.0	295	261/238	149	79/70
ZnS	Z – B	312	379	0.85	339	298	155	93
							or	or
							190	115
	H E X				346	318/297	181	92/73
ZnO	H E X	414	591	1.07	489	441/426	243	137/101
ZnTe	Z – B	190	210	0.53				
CdSe	H E X	186	218	0.71				
CdTe	Z – B	140	169	0.74				
BeO(E‖C)	H E X	680	1083	1.12	895	700/673	644	520/500
(E‖C)		724	1098	1.09				

Data for Table III is from Mitra (1963) (ZnSe), Marshall and Mitra (1964) (CdS, ZnS), Mitra and Gielisse (1964) (ZnO), Halsted *et al.* (1961) (ZnTe, CdSe, CdTe), and Loh (1968) (BeO).

2. *Group III–V Type Compounds*

The increased multiphonon structure in the semiconductor III–V compounds is extremely well illustrated by the study of AlSb by Turner and Reese (1962), who observed no less than forty-five multiphonon bands: seven two-phonon, ten three-phonon and twenty-eight four-phonon. The spectrum is well accounted for using the four-phonon scheme TO = 316, LO = 297, LA = 132 and TA = 65 cm^{-1}. However, the authors also subjected their data to a critical point analysis, producing the critical point energies shown in Table IV.

Gallium phosphide (Johnson 1965; Kleinman and Spitzer 1960), gallium arsenide (Johnson 1965; Cochran *et al.* 1961), and indium antimonide (Johnson 1965; Fray *et al.* 1960) have also been thoroughly investigated using critical point analysis. These results are also included in Table IV.

Data for Table IV is from Turner and Reese (1962) (AlSb), Kleinman and Spitzer (1960) (GaP), Cochran *et al.* (1961) (GaAs) and Fray *et al.* (1960) (InSb).

TABLE IV. Critical Point Frequencies (cm^{-1}) of Some Group III–V Type Compounds.

		AlSb	GaP	GaAs	InSb
$\Gamma(0,0,0)$	LO	339	392	286	206
	TO	317	367	270	190
$L(\frac{1}{2},\frac{1}{2},\frac{1}{2})$	TO	306	377	263	180
	LO	233	329	237	161
	LA	215	279	207	137
	TA	56	69	58	44
$X(1,0,0)$	TO	298	361	257	177
	LO	—	311	239	160
	LA	—	265	219	143
	TA	82	116	76	—
q^*/Ze		0.53	0.58	0.51	0.42
$\omega_t(k=0)$		318	366	273	185
$\omega_l(k=0)$		345	402	297	197

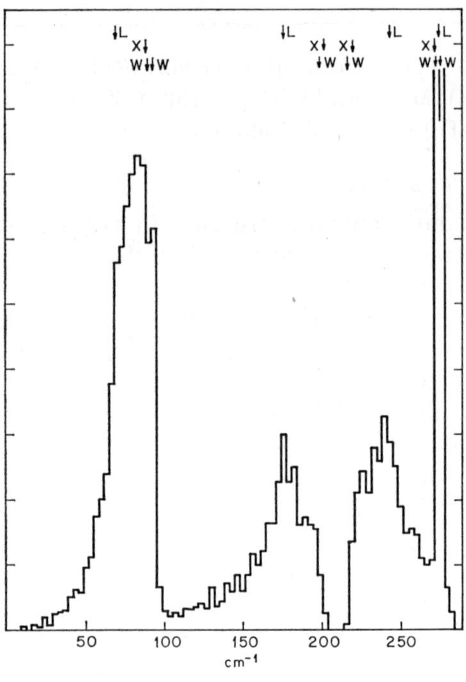

FIG. 15. The density of states for gallium arsenide.

In order to illustrate some of the points outlined in the theoretical section, gallium arsenide will be considered in some detail. Johnson and Cochran (1962) performed a shell model (nearest-neighbour) calculation producing the density-of-states curve and density-of-combined-states curves shown in Figs. 15 and 16, respectively.

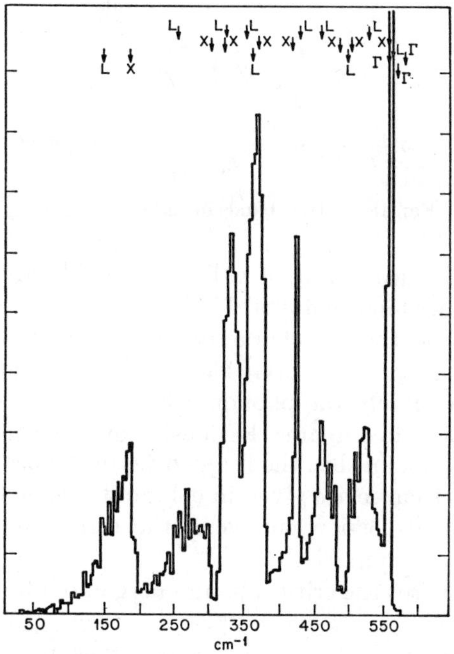

FIG. 16. The density of combined states for gallium arsenide.

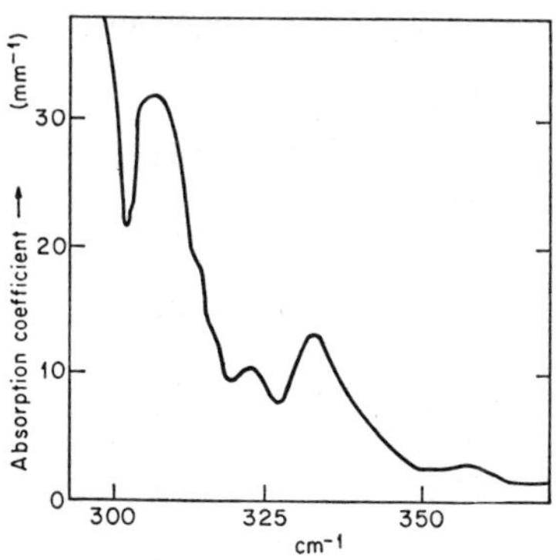

FIG. 17. Lattice bands in gallium arsenide.

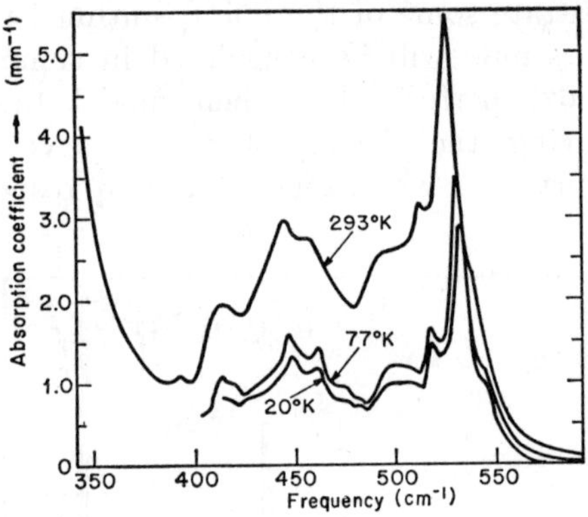

FIG. 18. Lattice bands in gallium arsenide.

The observed lattice bands of gallium arsenide are shown in Figs. 17 and 18; these are analysed in Table V.

The frequencies of the five characteristic zone boundary phonons are thus $TO_1 = 261$, $TO_2 = 255$, $LO = 234$, $LA = 186$, and $TA = 73$ cm^{-1}. To illustrate how closely the phonon spectrum follows the density-of-states curve, Table VI compares the density-of-states maxima for GaAs. As further illustration of this, the frequencies of the peaks in the density of combined states curve are given in column three of Table V.

The calculated dispersion curves in the directions $\Gamma \rightarrow L$, $\Gamma \rightarrow X$ and $X \rightarrow W$ are shown in Fig. 19.

Table VII compares the critical point energies obtained from the shell

TABLE V. Analysis of Lattice Bands of Gallium Arsenide (77°K).

Observed frequency (cm^{-1})	Assigned frequency (cm^{-1})	Calculated frequency (cm^{-1})	Assignment
523	522	553	$2TO_1$
509	510		$2TO_2$
494	495	519	$TO_1 + LO$
468	468		2LO
456	447	454	$TO_1 + LA$
442	441		$TO_2 + LA$
411	420	418	LO + LA
387?	372		2LA
333	334	361	$TO_1 + TA$
321	328		$TO_2 + TA$
306	307	325	LO + TA

TABLE VI. Comparison of Calculated and Observed Density-of-States Maxima for GaAs.

	Infrared (cm^{-1})	Calculated (cm^{-1})
TO	258	274
LO	234	242
LA	186	178
TA	73	89

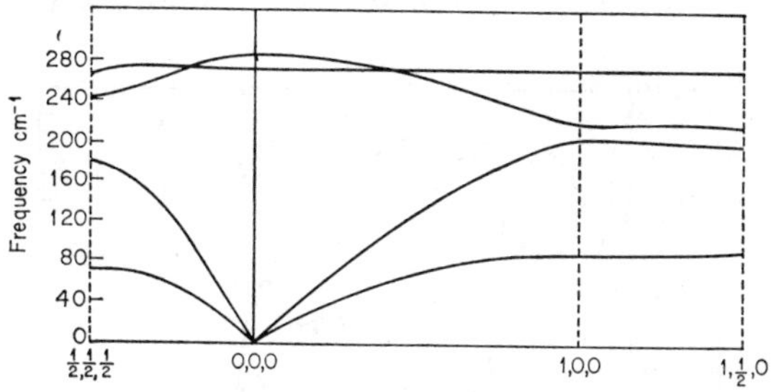

FIG. 19. Dispersion curves for gallium arsenide.

model calculation with those deduced from infrared and neutron scattering experiments. It can be seen that there is excellent agreement between results obtained from infrared and neutron scattering spectroscopy. It is also evident that the shell model (nearest-neighbour) calculation reasonably accounts for the experimental results but is not

TABLE VII. Comparison of Critical Point Frequencies of Gallium Arsenide at 77°K.

		Shell Model (cm^{-1})	Infrared (cm^{-1})	Neutron (cm^{-1})
Γ	LO	286	286	285
	TO	274	270	269
L	TO	277	263	261
	LO	245	237	237
	LA	178	207	209
	TA	71	58	62
X	TO	274	257	252
	LO	222	239	241
	LA	205	219	227
	TA	90	76	79

accurate in detail. Deviations of up to 15% are encountered between the observed and calculated dispersion curves.

Significant improvement may occur with the inclusion of second-neighbour forces (Dolling 1962). Fig. 19 and Tables VI and VII clearly indicate this discrepancy and show that the shell model tends to make transverse branches too isotropic and the transverse optic branches, in particular, too flat.

3. *Group IV–IV Type Compounds*

Multiphonon absorptions have been observed for the diamond-type semiconductors diamond, silicon and germanium (as well as for the

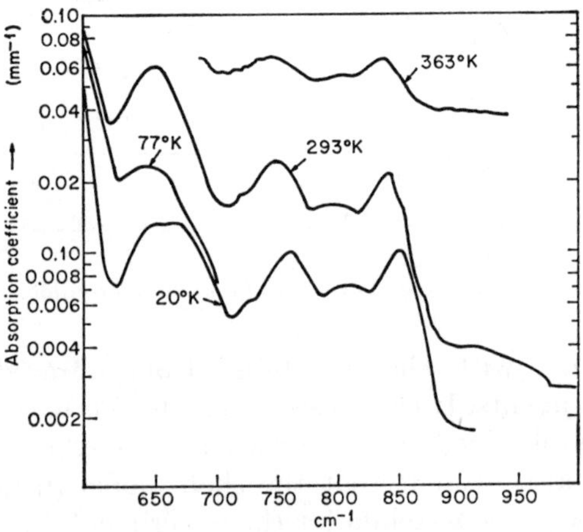

FIG. 20. Lattice bands in germanium.

heteropolar covalent compound silicon carbide). Diamond has been investigated using both infrared and Raman spectroscopy; silicon and germanium by both infrared and neutron scattering spectroscopy.

Whilst each of the above compounds has been treated in considerable detail (that is, shell model calculation and critical point analysis), in order to avoid repetition, germanium will be used for purposes of illustration and equivalent results for other compounds merely quoted.

The results of the infrared study of germanium by Fray *et al.* (1965) are shown in Figs. 20 and 21. It can be seen that the absorption falls into two distinct regions: one of relatively strong bands below about 610 cm^{-1}, and a much weaker region between approximately 610–890 cm^{-1}. The extremely sharp decrease in absorption at 610 cm^{-1} is known as the two-phonon "threshold" or "cut-off" and this, (a) marks the upper frequency limit for two-phonon absorption, and (b) provides an accurate measure of $2\omega_t$, where ω_t is the transverse optic phonon frequency.

It may be noted, however, that the absorption in this typical group IV–IV type compound is much weaker than that in the group III–V type compounds.

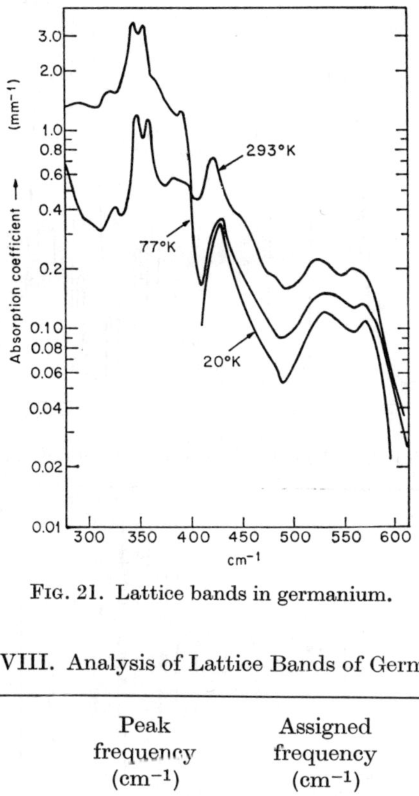

FIG. 21. Lattice bands in germanium.

TABLE VIII. Analysis of Lattice Bands of Germanium.

Assignment	Peak frequency (cm^{-1})	Assigned frequency (cm^{-1})	Calculated frequency (cm^{-1})
3TO	846	845	
2TO + LO	800	804	
2TO + LA	750	746	
3LO	723	724	
2TO	563	563	568
TO + LO	523	523	531
TO + LA	453	464	470
LO + LA	424	423	428
TO + TA	347	347	357
LO + TA	315	307	323

TO = 282, LO = 241, LA = 182, TA = 65 cm^{-1}.

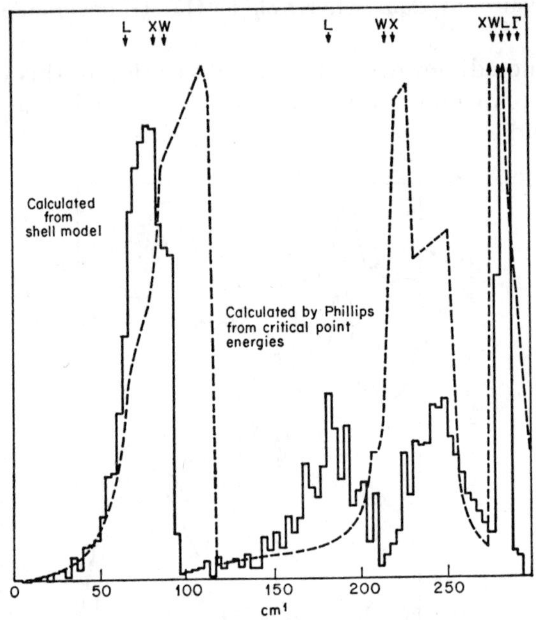

FIG. 22. The density of states for germanium.

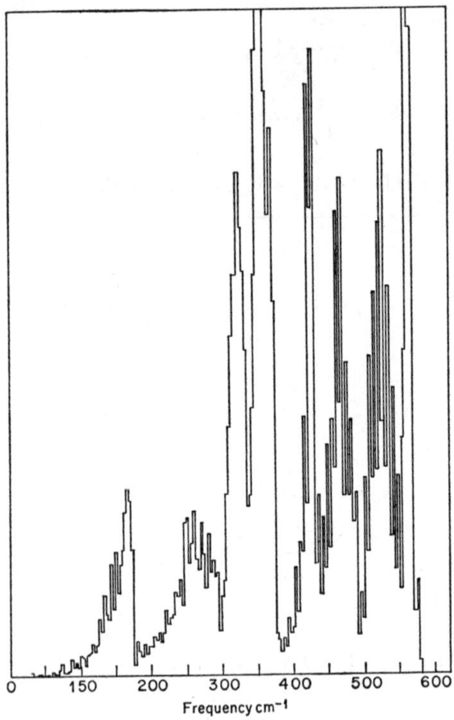

FIG. 23. The density of combined states for germanium.

Initially the data were analysed using a four-phonon scheme (that is, assuming the density-of-states curve contained four peaks) producing the results given in Table VIII. A nearest-neighbour shell model calculation was used to evaluate a density-of-states curve (Fig. 22), a density-of-combined-states curve (Fig. 23) and dispersion curves in the directions $\Gamma \rightarrow$ L, $\Gamma \rightarrow$ X and X $\rightarrow$ W (Fig. 24).

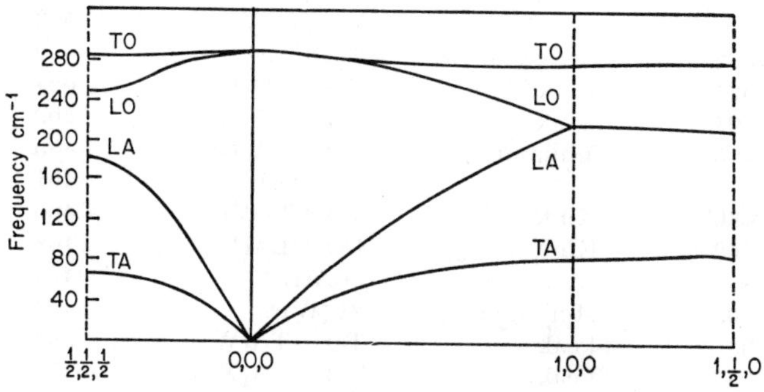

FIG. 24. Dispersion curves for germanium.

Table IX shows the reasonable agreement between the density-of-states maxima calculated and those obtained from the infrared.

TABLE IX. Density-of-States Maxima for Germanium.

	Infrared (cm^{-1})	Calculated (cm^{-1})
TO	282	282
LO	241	246
LA	182	181
TA	65	77

The frequencies of the maxima in the calculated density-of-combined-states curve are included in column four of Table VIII for comparison with the observed peaks.

Fray *et al.* (1965, *vide supra*) then identified features in the absorption curve associated with critical points at Γ, L, X and W (see Fig. 8). The results of this critical point analysis are given in Table X. Table XI lists the critical point energies of germanium, silicon and diamond, obtained from either infrared (IR), Raman (R), neutron scattering (N) or shell model calculation (Calc).

TABLE X. Critical Point Frequencies for Germanium.

Observed frequency (cm^{-1})	Feature	Assignment	Calculated frequency (cm^{-1})
610	Minimum	2O[a]	610
555	Kink	2TO(L)[a]	557
		or 2TO(W)[a]	553
532	Kink	2TO(X)[a]	539
524	Peak	TO + LO(L)	524
504	Kink	TO + LA(X)	502
495	Minimum	TO + LA(L)[a]	495
		or 2LO(L)[a]	491
480	Kink	TO + L(W)	480
462	Kink	LO + LA(L)	462
436	Kink	2LA(L)[a]	434
405	Minimum	2L(W)[a]	407
388	Peak	TO + TA(W)	388
353	Peak	TO + TA(X)	352
342	Peak	TO + TA(L)	342
		L + TA(X)	315
		or L + TA(W)	315
308	Minimum	LO + TA(L)[a]	309

[a] Forbidden combination.

TABLE XI. Critical Point Frequencies (cm^{-1}) of Some IV–IV Type Compounds.

Critical point		Germanium IR	Germanium N	Germanium Calc	Silicon IR	Silicon N	Silicon Calc	Diamond IR	Diamond R
Γ	O	305	300	291	520	518	498	1335	1333
L	TO	278	247	249	418	420	449	1162	1167
	LA	217	215	183	377	378	294	992	996
	TA	64	64	68	119	115	106	508	—
X	TO	269	275	279	459	464	494	1234	1235
	L	232	230	219	407	411	378	1121	1128
	TA	82	81	84	155	150	133	847	—
W	TO	277	—	282	488	—	499	1275	1251
	L	203	—	214	370	—	370	1057	1095
	TA	111	—	89	169	—	136	936	925

The zone-centre phonon frequencies, effective charge and zone-boundary phonon frequencies, based on a four-phonon scheme, are summarized in Table XII for the above compounds.

It may be noticed that in the above table the silicon carbide LO phonon frequency has been assigned a higher value than the corresponding TO phonon frequency, while the reverse obtains for diamond, silicon and germanium. Keyes (1962) noted a correlation between the effective

TABLE XII. Zone-Centre and Zone-Boundary Phonon Frequencies (cm^{-1}) of Some Group IV–IV Type Compounds.

	ω_t	ω_l	q^*/Ze	TO	LO	LA	TA
Diamond	1335	1335	0	1275	1161	991	750
Silicon	520	520	0	482	413	333	139
Germanium	305	305	0	282	241	182	65
SiC	793	970	0.94	770	851	540	363

charge q^*/Ze of some zinc-blende type crystals with their zone-boundary phonon frequencies. For homopolar crystals ($q^* = 0$) the two optical branches are degenerate at the centre of the Brillouin zone, but Keyes found that the frequency of the longitudinal optical branch increased

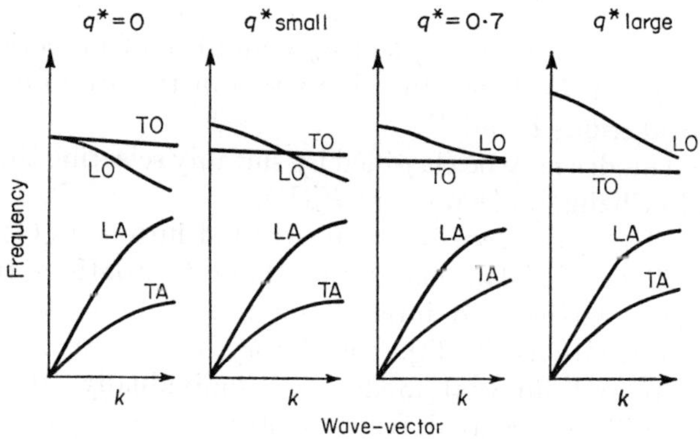

FIG. 25. The effect of effective charge q^* on the optical branches of the dispersion diagram.

relative to the transverse optic branch, as the effective charge increased. His results predicted that for an effective charge of 0.7, the two optic branches were degenerate at the zone boundary. Mitra (1963) has shown this to hold for zinc selenide. The qualitative effect of q^* on the dispersion curves is shown in Fig. 25.

D. Molecular Crystals

For a crystal involving molecular groupings, two methods have been employed to determine the irreducible representations involved, and the number of unit-cell fundamentals (together with infrared and Raman activity) contained in each of these representations (Bhagavantam and Venkatarayudu 1951; Halford 1946). Hornig (1948) has indicated that both methods produce equivalent results.

Shimanouchi *et al.* (1961) have adapted the well-known GF matrix method, widely used to calculate vibration frequencies of a molecule, to evaluate the optically active lattice vibration frequencies of a crystal.

The procedure of Bhagavantam and Venkatarayudu (1951) is outlined and applied to a molecular crystal in a following example which also indicates that if the space group of a crystal is not uniquely known, a study of the crystal lattice vibrations may help to resolve the difficulty.

The method consists of determining the symmetry operations of the unit cell and then writing out the character table (these already appear as an appendix in several spectroscopic texts). The number of times (n_i) a particular irreducible representation Γ_i is contained in another representation, Γ, is given by

$$n_i = \frac{1}{N} \sum h_\rho \, \Psi'_\rho(\mathrm{R}) \, \Psi_i(\mathrm{R}) \tag{14}$$

where $N = \sum h_\rho =$ the group order; $h_\rho =$ number of operations in the class ρ; $\Psi'_\rho(\mathrm{R})$ and $\Psi_i(\mathrm{R})$ are the characters of the group operation R in the representations Γ and Γ_i.

The normal modes may be classified by suitably selecting the representation Γ and utilizing its characters $\Psi'_\rho(\mathrm{R})$.

If $\Psi'_\rho(\mathrm{R}) = U_\mathrm{R}\ (\pm 1 + 2\cos\phi_\mathrm{R})$ is substituted into eqn (14), then the resulting n_i will include all normal modes, that is translations, translatory lattice, rotatory lattice and internal modes. In order to obtain the number of translations $[n_i(\mathrm{T})]$, translatory lattice $[n_i(\mathrm{T}')]$, rotatory lattice $[n_i(\mathrm{R}')]$ and internal modes (n'_i) individually, the following expressions for $\Psi'_\rho(\mathrm{R})$ are used, respectively:

$$\Psi'_\rho(\mathrm{R}) = \pm 1 + 2\cos\phi_\mathrm{R}\colon\ n_i(\mathrm{T})$$
$$\Psi'_\rho(\mathrm{R}) = [U_\mathrm{R}(s) - 1]\,(\pm 1 + 2\cos\phi_\mathrm{R})\colon\ n_i(\mathrm{T}')$$
$$\Psi'_\rho(\mathrm{R}) = [U_\mathrm{R}(s - v)]\,(1 \pm 2\cos\phi_\mathrm{R})\colon\ n_i(\mathrm{R}')$$
$$\Psi'_\rho(\mathrm{R}) = [U_\mathrm{R} - U_\mathrm{R}(s)]\,(\pm 1 + 2\cos\phi_\mathrm{R}) - U_\mathrm{R}(s - v)\,(1 \pm 2\cos\phi_\mathrm{R})\colon\ n'_i$$

where U_R is the number of atoms in a unit cell invariant under the operation R; $U_\mathrm{R}(s)$ is the number of structural groups remaining invariant under the operation R; $U_R(s-v)$ is the number of polyatomic groups remaining invariant under the operation R; ϕ_R is the angle of rotation corresponding to operation R. The positive sign in the above expressions is used when R is a proper rotation and the negative sign when R is an improper rotation. In this connexion the identity operation, E, is equivalent to a proper rotation through 0°; the inversion operation, i, is equivalent to an improper rotation through 180°; and the reflection operation, σ, is equivalent to an improper rotation through 0°.

The molecular crystal hydrazine, N_2H_4, will now be considered. The crystal structure has been determined as C_2^2 by Busing *et al.* (1961) and C_{2h}^2 by Collin and Lipscomb (1951). Written out below are the character tables and results of the above procedure applied to the point groups corresponding to the space groups C_2^2 and C_{2h}^2.

C_2	E	C_2	T	T′	R′	IR	R
a	1	1	1	2	3	√	√
b	1	−1	2	1	3	√	√
$U_R(s)$	2	0	$N = 2$	$s = 2$			
$U_R(s-v)$	2	0	$v = 0$				
$h_\rho \Psi'_\rho(\mathrm{T})$	3	−1					
$h_\rho \Psi'_\rho(\mathrm{T}')$	3	1					
$h_\rho \Psi'_\rho(\mathrm{R}')$	6	0					

C_{2h}	E	C_2	σ_h	i	T	T′	R′	IR	R
a_g	1	1	1	1	0	2	1		√
a_u	1	1	−1	−1	1	0	2	√	
b_g	1	−1	−1	1	0	1	2		√
b_u	1	−1	1	−1	2	0	1	√	
$U_R(s)$	2	0	2	0					
$U_R(s-v)$	2	0	2	0		$N = 4$	$s = 2$		
$h_\rho \Psi'_\rho(\mathrm{T})$	3	−1	1	−3		$v = 0$			
$h_\rho \Psi'_\rho(\mathrm{T}')$	3	1	1	3					
$h_\rho \Psi'_\rho(\mathrm{R}')$	6	0	−2	0					

As far as lattice modes are concerned, not more than nine bands in the infrared are expected if the crystal structure is C_2^2 and not more than three bands if the crystal structure is C_{2h}^2.

Baglin, Bush and Durig (1967) examined the far infrared spectra of hydrazine and hydrazine-d_4 and obtained the results shown in Figs. 26 and 27.

The observed number of lattice modes (six for hydrazine and seven for hydrazine-d_4) clearly indicates the crystal structure to be C_{2h}^2. Very few lattice overtone or combination bands of appropriate intensity have

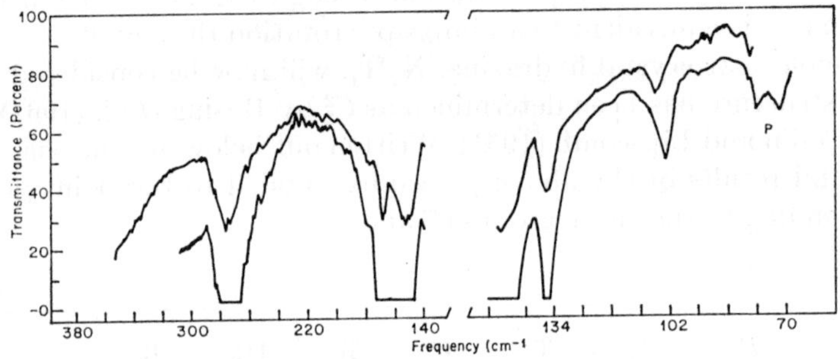

FIG. 26. The far-infrared spectrum of crystalline hydrazine. *P* indicates bands due to polyethylene.

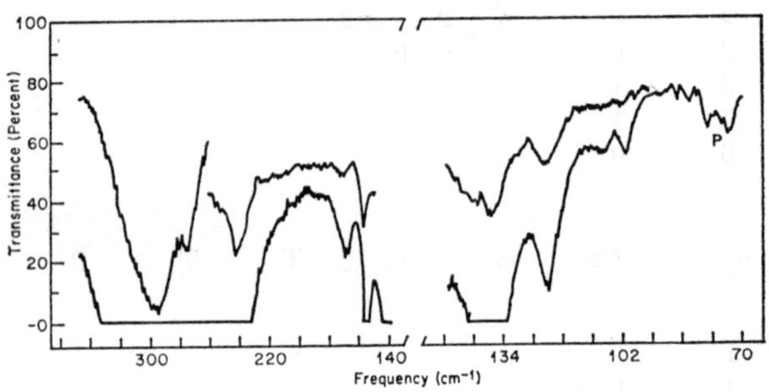

FIG. 27. The far infrared spectrum of crystalline hydrazine-d_4. *P* indicates bands due to polyethylene.

been observed experimentally and the detection of some three or four such bands in a single spectrum may be considered extremely improbable.

The authors' assignment of the various lattice modes is given in Table XIII.

The crystal spectrum of yttrium vanadate YVO_4 (Miller *et al.* 1968) may be considered in order to illustrate a complete group theoretical analysis. The space group of YVO_4 is D_{4h}^{19} and the crystal has two molecules per unit cell. The results of a Bhagavantam and Venkatarayudu-type procedure are shown in Table XIV.

TABLE XIII. Assignment of Lattice Modes of N_2H_4 and N_2D_4.

N_2H_4 (cm^{-1})	274.9	180.5	172.5	154.0		136.3	106.7
N_2D_4 (cm^{-1})	243.0	169.5	156.5	142.3	136.9	123.0	101.5
Type	L	T(A)	L	T(A)	L	L	T(B)

L = libratory, T = translatory.

TABLE XIV

D_{4h}	E	$2C_4$	C_2	$2C_2'$	$2C_2''$	i	$2S_4$	σ_h	$2\sigma_v$	$2\sigma_d$	n_i	T	T′	R′	n_i'	IR	R
a_{1g}	1	1	1	1	1	1	1	1	1	1	2	0	0	0	2		√
a_{2g}	1	1	1	−1	−1	1	1	1	−1	−1	1	0	0	1	0		
b_{1g}	1	−1	1	1	−1	1	−1	1	1	−1	1	0	0	0	1		√
b_{2g}	1	−1	1	−1	1	1	−1	1	−1	1	4	0	2	0	2		√
e_g	2	0	−2	0	0	2	0	−2	0	0	5	0	2	1	2		√
a_{1u}	1	1	1	1	1	−1	−1	−1	−1	−1	1	0	0	0	1		
a_{2u}	1	1	1	−1	−1	−1	−1	−1	1	1	4	1	1	0	2	√	
b_{1u}	1	−1	1	1	−1	−1	1	−1	−1	1	2	0	0	0	2		
b_{2u}	1	−1	1	−1	1	−1	1	−1	1	−1	1	0	0	1	0		
e_u	2	0	−2	0	0	−2	0	2	0	0	5	1	1	1	2	√	
U_R	12	0	4	4	0	0	4	0	0	8							
$U_R(s)$	4	0	4	4	0	0	4	0	0	4							
$U_R(s-v)$	2	0	2	2	0	0	2	0	0	2	$N = 16$			$s = 4$			
$h_\rho \Psi_\rho'(n_i)$	36	0	−4	−8	0	0	−8	0	0	16		$v = 2$					
$h_\rho \Psi_\rho'(\mathrm{T})$	3	2	−1	−2	−2	−3	−2	1	2	2							
$h_\rho \Psi_\rho'(\mathrm{T'})$	9	−2	−3	−6	2	3	−6	−1	−2	6							
$h_\rho \Psi_\rho'(\mathrm{R'})$	6	0	−2	−4	0	0	4	0	0	−4							
$h_\rho \Psi_\rho'(n_i')$	18	0	2	4	0	0	−4	0	0	12							

Since the unit cell contains twelve atoms, thirty-six modes of vibration are expected; the results above indicate that these are distributed among the following irreducible representations:

$$\Gamma_{36}(D_{4h}) = 2a_{1g} + a_{2g} + b_{1g} + 4b_{2g} + 5e_g + a_{1u} + 4a_{2u} + 2b_{1u} + b_{2u} + 5e_u.$$

The above results also predict that, barring accidental degeneracies, the infrared spectrum should consist of seven bands associated with four internal, two translatory lattice, and one rotatory lattice modes, while twelve bands should appear in the Raman spectrum, arising from seven internal, four translatory lattice and one rotatory lattice modes.

In order to aid assignment, it is useful to consider the crystal to be made up of separate entities: Y^{3+} and VO_4^{3-}. The vanadate ion has tetrahedral symmetry (T_d) and the fifteen expected modes are found to take the following representation: $\Gamma_{15}(T_d) = a_1 + e + t_1 + 3t_2$ where t_1 is equivalent to pure rotation and one of the t_2's corresponds to a pure translation. Thus, the infrared spectrum of the vanadate ion, in solution, would be expected to consist of four bands, and these have been determined for aqueous Na_3VO_4 by Siebert (1954) to be $\nu_1(a) = 870$, $\nu_2(e) = 345$, $\nu_3(t_2) = 825$ and $\nu_4(t_2) = 480\ cm^{-1}$.

Now in the crystal of YVO_4, the vanadate (and yttrium) ions possess D_{2d} site symmetry and by constructing a correlation chart between T_d and D_{2d} point groups (Table XV), the expected mode pattern arising from the two vanadate ions per unit cell may be written as follows: $\Gamma_{30}(D_{2d}) = 4a_1 + 2b_1 + 2a_2 + 6b_2 + 8e$.

TABLE XV. Correlation Chart for Point Groups T_d and D_{2d}.

T_d	D_{2d}
a_1	a_1
a_2	b_1
e	$a_1 + b_1$
f_1	$a_2 + e$
f_2	$b_2 + e$

TABLE XVI. Correlation Chart for Point Groups D_{2d} and D_{4h}.

D_{2d}	D_{4h}
a_1	$a_{1g} + b_{1u}$
b_1	$b_{1g} + a_{1u}$
a_2	$a_{2g} + b_{2u}$
b_2	$b_{2g} + a_{2u}$
e	$e_g + e_u$

The z coordinate of an yttrium ion transforms as the symmetry species b_2 of the point group D_{2d}, while the x and y coordinates transform as E.

Thus, the yttrium ions contribute six modes $\Gamma_6(D_{2d}) = 2b_2 + 2e$. The total representation is therefore $\Gamma_{36}(D_{2d}) = 4a_1 + 2b_1 + 2a_2 + 8b_2 + 10e$. Reference to the correlation chart between point groups D_{2d} and D_{4h} (Table XVI) enables the above representation to be written as:

$$\Gamma_{36}(D_{4h}) = (2a_{1g} + 2b_{1u}) + (b_{1g} + a_{1u}) + (a_{2g} + b_{2u}) + (4b_{2g} + 4a_{2u}) + (5e_g + 5e_u)$$

which is exactly equivalent to the expression derived earlier from the procedure of Bhagavantam and Venkatarayudu.

A correlation chart for the symmetry species of the T_d group of the free VO_4^{3-} ion, the D_{2d} group of the VO_4^{3-} and Y^{3+} sites in the crystal, and the D_{4h} group of the crystal itself follows.

Inter-correlation Chart for the Various Point Groups Involved in Crystalline YVO_4

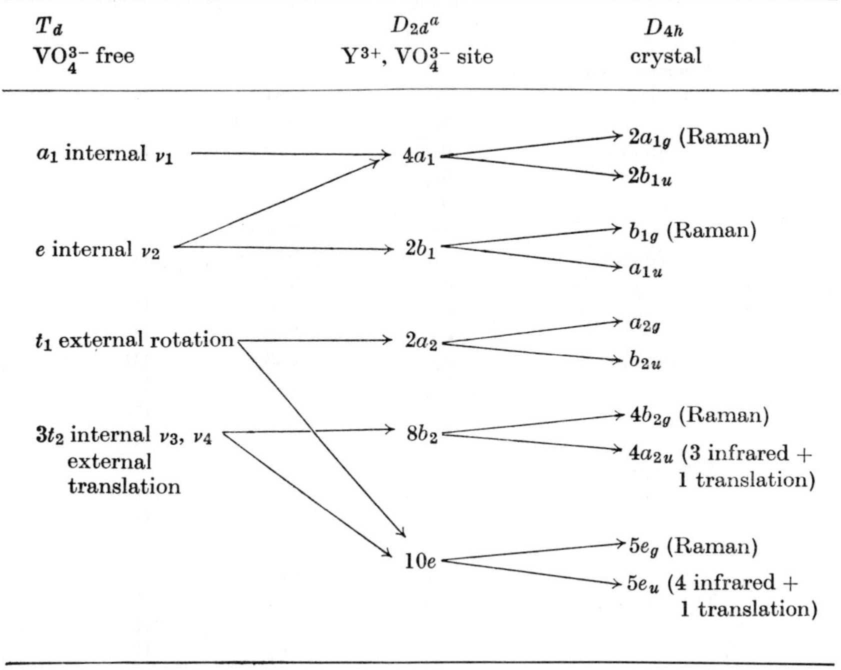

[a] An yttrium ion contributes ($b_2 + e$). The number of modes is doubled due to the unit cell containing two molecules.

Miller *et al.* (1968) obtained Raman frequency and polarization data on YVO_4 and also measured the infrared emittance spectrum. They were able to assign six of the seven infrared active modes, and nine of the twelve Raman active modes to features in their spectra. Table XVII summarizes their assignments.

E. Defect-induced Lattice Spectra

The relatively modern technique of introducing defects, either physically or chemically, into a crystal which, when pure, exhibits no one-phonon absorption, has proved to be extremely useful in providing information on the phonon frequency spectrum of the pure host lattice, and on the inter-atomic forces operative around the defect.

Band mode absorption was observed as long ago as 1934 by Robertson *et al.* (1934) for type 1 diamond; in the last decade, spectra of diamond, silicon and germanium have been recorded using neutron irradiation to

TABLE XVII. Assignment of the Crystal Spectra of YVO_4.

Mode type	Species	Frequency (cm^{-1})	
Internal modes			
ν_1 (870 cm^{-1})	a_{1g}	891	R
	b_{1u}	inactive	
ν_2 (345 cm^{-1})	a_{1g}	379	R
	b_{1u}	inactive	
	b_{1g}	260	R
	a_{1u}	inactive	
ν_3 (825 cm^{-1})	b_{2g}	817	R
	a_{2u}	803	IR
	e_g	840	R
	e_u	788	IR
ν_4 (480 cm^{-1})	a_{2u}	455	IR
	b_{2g}	489	R
	e_g	—	(R)
	e_u	—	(IR)
External modes			
Rotation	e_g	—	(R)
	e_u	261	IR
	a_{2g}	inactive	
	b_{2u}	inactive	
Translation	b_{2g}	157	R
	b_{2g}	—	(R)
	b_g	162	R
	b_g	260	R
	a_{2u}	310	IR
	e_u	196	IR

produce the defects. However, the nature of the defects was not known in detail in the above crystals and, as mentioned earlier, in order to test theoretical predictions, attention should be confined to systems where the form of the defect is known.

The absorption of boron and phosphorus in a silicon host lattice was observed by Angress *et al.* (1965). Two "doping" elements were used to avoid free carrier absorption, which would otherwise mask any lattice effects. The spectra obtained are shown in Fig. 28.

In agreement with what has been said in Section I on defect-induced lattice spectra, localized modes due to the boron (lighter than silicon) but none due to phosphorus (heavier than silicon) are expected, while both impurities will "activate" band modes of the silicon lattice. The localized modes should appear sharp and at frequencies greater than the highest allowed one-phonon frequency of the lattice (determined from

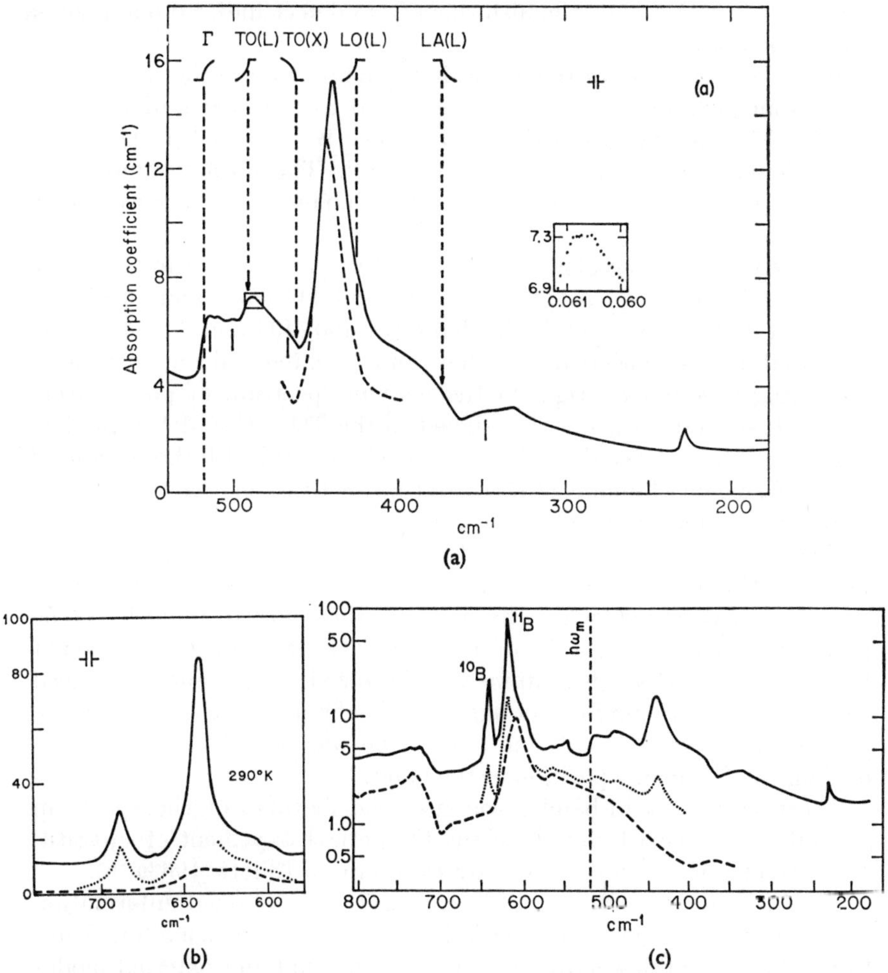

FIG. 28. (a) Band mode absorption of silicon doped with both boron and phosphorus in concentrations of 5×10^{19} cm^{-3}. —— 290°K, --- 80°K. Inset are the experimental points over the TO region. (b) Boron (concentration 5×10^{19} cm^{-3}) activated fundamental local mode absorption bands at 290°K after different doses of compensating 2 MeV electrons. —— after 4.7×10^{18} electrons cm^{-2}, · · · after 9.4×10^{18} electrons cm^{-2}, --- pure silicon. (c) The infrared spectrum of silicon doped with both boron and phosphurus in concentrations of 5×10^{19} cm^{-3} (——) and 5×10^{18} cm^{-3} (· · ·). The spectrum of pure silicon (---) is shown for comparison.

Raman spectroscopy to be 520 cm^{-1} for silicon). The non-localized band modes should therefore appear in the frequency range below 520 cm^{-1}.

The spectrum of pure silicon (also shown in Fig. 28) contains two two-phonon absorption peaks and these are labelled TO + TA and

LO + TA in order to distinguish them from defect-induced one-phonon absorption peaks.

The major features of the defect-induced spectra are the sharp peaks at 644 and 620 cm^{-1}, which were assigned to the modes of vibration localized in the vicinity of the boron atoms B^{10} and B^{11}, respectively. (Calculated positions were 660 and 635 cm^{-1}.) The authors showed that the above absorptions were characteristic of unaggregated substitutional boron.

In the band mode region, five main features are evident in the doped samples—three sharp bands at 516, 490 and 441 cm^{-1} and two broad bands at 403 and 335 cm^{-1}. The bands at 490, 403 and 335 cm^{-1} also appeared in the spectrum of neutron-irradiated silicon specimens, indicating a common origin in the phonon spectrum of pure silicon. These three frequencies were assigned to the TO, LO and LA maxima in the density of states since their frequencies agreed with those deduced by Johnson and Loudon (1964) from an analysis of the two-phonon infrared combination bands.

The most striking feature of the band mode region is the strong, sharp band at 441 cm^{-1}. The authors concluded that this was due to an in-band resonance associated with the heavy (phosphorus) impurity. The parameters required for a theoretical fit to 441 cm^{-1}, however, were not those for single phosphorus atoms with localized charges and unchanged force constants. Some form of aggregation of the phosphorus atoms to produce an effective oscillating mass of 66 was postulated in order to explain the frequency position of the band.

Apart from this in-band resonance, the features of the spectrum generally agreed with details of the theoretical treatment of substitutional impurities in silicon given by Dawber and Elliott (1963a, b).

Very recently Cundill and Sherman (1968) obtained information concerning lattice modes of a crystal using a method which is analogous to observing combination bands between internal and external modes in a pure crystal. Initially they investigated the $\nu_3 \pm \nu_{ext}$ spectrum of the NCO^- ion isolated in KBr and the $\nu_1 \pm \nu_{ext}$ spectrum of the CN^- ion, also isolated in KBr. The spectra obtained are shown in Figs. 29 and 30. Dealing first with the case of NCO^-, the external local mode frequencies of various isotopic species of the NCO^- ion were obtained [from combinations with ν_3 (NCO^-)], and from the resulting pattern the authors deduced that the four bands observed in any single NCO^- isotopic species all arose from motions which were predominantly torsional. The motions are depicted in Fig. 31. The terms "acoustic" and "optic" refer only to the three nearest neighbours to the large amplitude nucleus of the NCO^- ion.

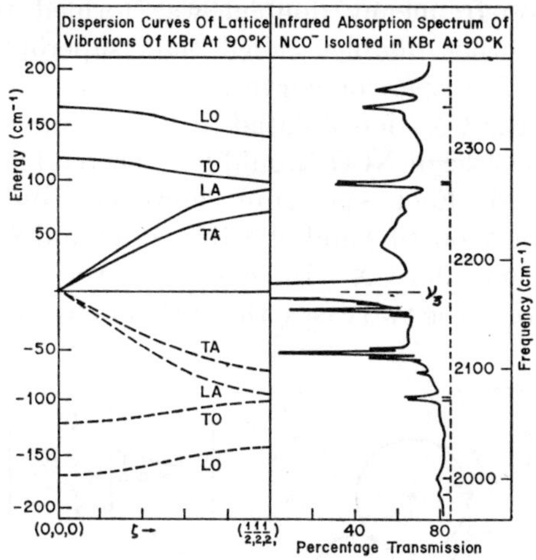

FIG. 29. The absorption spectrum of NCO^- isolated in KBr, showing the external mode vibrations in sum and difference combinations with ν_3 compared with the (1, 1, 1) dispersion curves of the lattice vibrations of pure KBr.

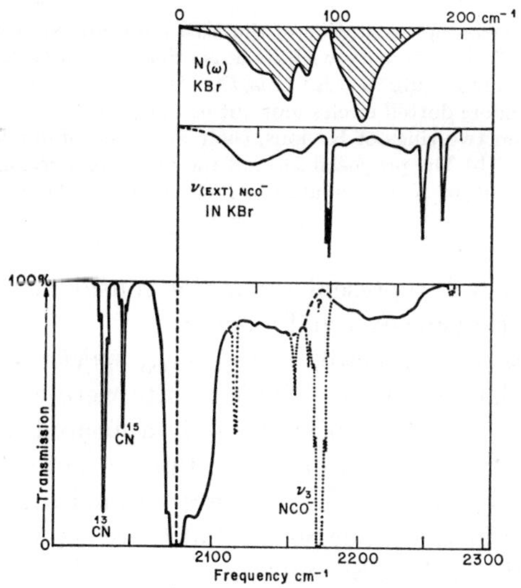

FIG. 30. The absorption spectrum of CN^--doped KBr at 7°K. The external modes seen in combination with the C–N stretching mode (2078 cm^{-1}) of the $(C^{12}N^{14})^-$ species are compared with the similar spectrum for NCO^- in KBr and the integrated density of states for KBr. Features due to NCO^- in the CN^--doped crystal have been shown dotted.

The sharp lower-frequency doublet was termed the "in-the-gap" doublet and the other doublet called the "superoptic" doublet, by reference to the KBr dispersion curves.

In the case of the CN^- ion isolated in KBr, none of the four sharp bands found in the case of NCO^- would be expected to occur and indeed none was observed. Each spectrum, however, contains two broad maxima within the acoustic band of KBr—at 44 and 80 cm^{-1} for NCO^- and at 12 and 75 cm^{-1} for CN^-. In considering the reason for the shifts in frequency between the two cases, the authors postulated that orienta-

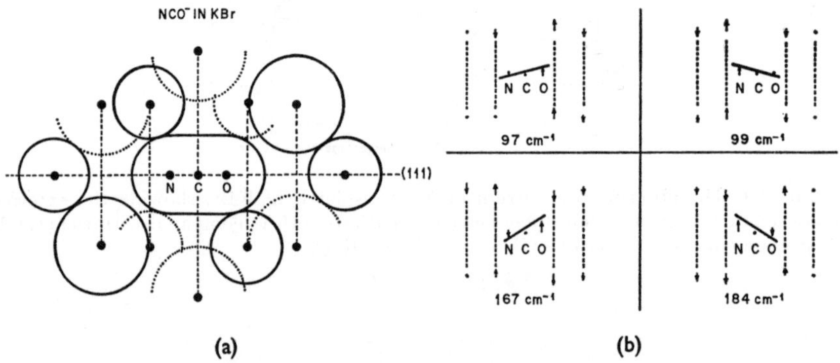

FIG. 31. (a) Schematic representation of the KBr environment surrounding an NCO^- present as a substitutional impurity with its axis oriented in a (1, 1, 1) direction: drawn to scale using Goldschmidt radii. o = K^+ ions, O = Br^- ions. Solid circles represent ions in the plane of the paper; dotted circles ions out of this plane. Note particularly how the NCO^- is held between two rings of K^+ ions, each composed of one ion in the plane and two out of this plane. (b) The proposed form of the vibrations giving sharp bands in the external mode spectrum of NCO^- isolated in KBr. ---- the planes of nearest and next nearest neighbours.

tion effects might well represent the major consideration for these low-frequency, acoustic-type external modes.

The authors also investigated the $\nu_3 \pm \nu_{ext}$ spectra for the NCO^- ion in twelve alkali halide lattices. Their results were consistent with the following facts. (a) In-the-gap modes only appeared if the host lattice contained a gap in its density of states. Furthermore, the separation of the in-the-gap doublet seemed to be proportional to the size of such a gap. The authors therefore suggested that there is no gap in the density of states for RbBr, RbI, CsBr and CsI; and also that the calculations of Karo and Hardy (1963) placed the gap too low in NaBr. (b) Superoptic modes appeared only when the nearest-neighbour ions were heavy enough to have small amplitudes in order to localize the motion and, hence, allow the frequency to clear the optic band of the host lattice.

One can thus appreciate that studies of this kind allow monitoring of the density of states for a host lattice.

The effects of temperature and pressure on the $\nu_{int} \pm \nu_{ext}$ spectra were also investigated. As the temperature is reduced and the lattice contracts, the lattice mode frequencies are expected to rise. Experimentally, it was found that the superoptic features narrowed and increased in separation

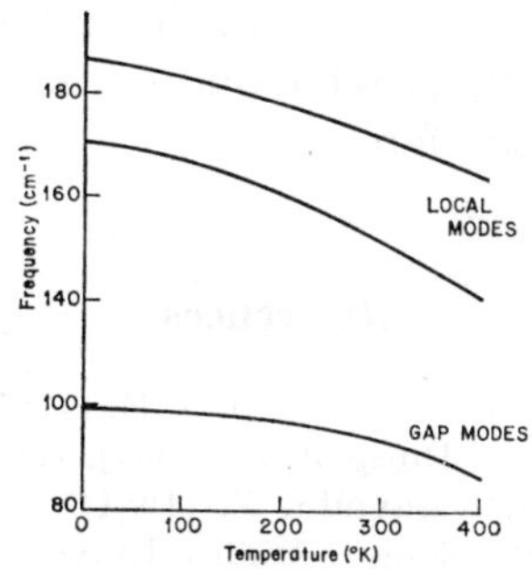

FIG. 32. The temperature dependence of the frequencies of the sharp external mode features found in combination with $\nu_3(NCO^-)$ in KBr.

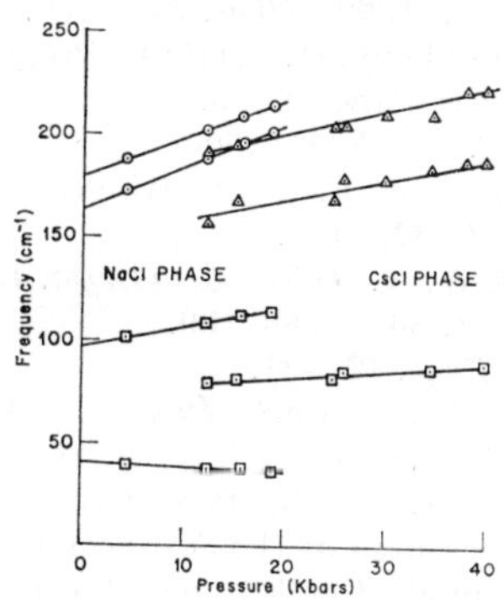

FIG. 33. The pressure dependence of the frequencies of the prominent features in the lattice-sideband structure on $\nu_3(NCO^-)$ isolated in KBr at 140°K.

from ν_{int} more rapidly than did the in-the-gap features: this is shown in Fig. 32. The effect of pressure on the system NCO^- isolated in KBr at 140°K is shown in Fig. 33. Here, the potassium halide changes phase from its low-pressure NaCl-type structure to its high-pressure CsCl-type structure. Investigation of the similar systems NCO in RbBr, RbI and KI resulted in the generalizations that at the phase change, (a) the superoptic doublet decreased in frequency, with an increase in doublet separation, (b) the broad low-frequency band was replaced by a sharper, higher-frequency band, and (c) no in-the-gap structure was observed in the high-pressure phase.

The unique behaviour of the lowest frequency band of the NaCl-type structure (decrease in the frequency with increase of pressure) might indicate the tendency towards instability as the phase change is approached. Generally, the steady frequency shifts (within a given phase) indicate the pressure dependence of the relevant force constant. When a band can be associated with a point in the host lattice dispersion curve, the frequency shifts provide information concerning the anharmonicity of the pure lattice forces.

References

Angress, J. F., Goodwin, A. R., and Smith, S. D. (1965). *Proc. R. Soc.* **A287**, 64.

Baglin, F. G., Bush, S. F., and Durig, J. R. (1967). *J. Chem. Phys.* **47**, 2104.

Bhagavantam, S. and Venkatarayudu, T. (1951). "Theory of Groups and its Application to Physical Problems". Andhra University, Waltair.

Birman, J. L. (1963). *Phys. Rev.* **131**, 1489.

Blackman, M. (1933). *Z. Phys.* **86**, 421.

Born, M. and von Karman, T. (1912). *Phys. Z.* **13**, 297.

Brout, R. and Visscher, W. M. (1962). *Phys. Rev. Lett.* **9**, 54.

Busing, W. R., Zocchi, M., and Levy, H. A. (1961). Program of the Annual Meeting of the American Cryst. Assn, Paper No. 3.

Cochran, W. (1959). *Proc. R. Soc.* **A253**, 260.

Cochran, W., Fray, S. J., Johnson, F. A., Quarrington, J. E., and Williams, N. J. (1961). *Appl. Phys.* (*Suppl.*) **32**, 2102.

Collin, R. L. and Lipscomb, W. N. (1951). *Acta Cryst.* **4**, 10.

Cowley, R. A. (1962). *Proc. R. Soc.* **A268**, 109.

Cowley, R. A. (1963). *Adv. Phys.* **12**, 421.

Cundill, M. A. and Sherman, W. F. (1968). *Phys. Rev.* **168**, 1007.

Dawber, P. G. and Elliott, R. J. (1963a). *Proc. R. Soc.* **A273**, 222.

Dawber, P. G. and Elliott, R. J. (1963b). *Proc. Phys. Soc.* (Lond.) **81**, 453.

Dolling, G. (1962). Paper presented at the Symposium on Inelastic Scattering of Neutrons in Solids and Liquids, Chalk River, Ontario.

Fray, S. J., Johnson, F. A., and Jones, R. (1960). *Proc. Phys. Soc. Lond.* **76**, 939.

Fray, S. J., Johnson, F. A., Quarrington, J. E., and Williams, N. J. (1965). *Proc. Phys. Soc. Lond.* **85**, 153.

Hass, M. (1960). *Phys. Rev.* **117**, 1497.

Haas, C. and Ketelaar, J. A. A. (1956). *Phys. Rev.* **103**, 564.

Halford, R. S. (1946). *J. Chem. Phys.* **14**, 8.

Halsted, R. E., Lorenz, M. R., and Segall, B. (1961). *J. Phys. Chem. Solids* **22**, 109.

Hornig, D. F. (1948). *J. Chem. Phys.* **16**, 1063.

Huang, K. (1951). *Proc. R. Soc.* (Lond.) **A208**, 352.

Johnson, F. A. (1965). *Progr Semicond* **9**, 181.

Johnson, F. A. and Cochran, W. (1962). Proc. Intern. Conf. on the Physics of Semiconductors, Exeter.

Johnson, F. A. and Loudon, R. (1964). *Proc. R. Soc.* **A281**, 274.

Jones, G. O., Martin, D. H., Mawer, P. A., and Perry, C. H. (1961). *Proc. R. Soc.* **A261**, 10.
Karo, A. M. and Hardy, J. R. (1963). *Phys. Rev.* **129**, 2004.
Keyes, R. W. (1962). *J. Chem. Phys.* **37**, 72.
Kleinman, D. A. (1960). *Phys. Rev.* **118**, 118.
Kleinman, D. A. and Spitzer, W. G. (1960). *Phys. Rev.* **118**, 110.
Lax, M. and Burstein, E. (1955). *Phys. Rev.* **97**, 39.
Loh, E. (1968). *Phys. Rev.* **166**, 673.
Loudon, R. (1964). *Proc. Phys. Soc. Lond.* **84**, 379.
Lyddane, R. H., Sachs, R. G., and Teller, E. (1941). *Phys. Rev.* **59**, 673.
Marshall, R. and Mitra, S. S. (1964). *Phys. Rev.* **134**, A1019.
Martin, D. H. (1965). *Adv. Phys.* **14**, 39.
Miller, S. A., Caspers, H. H., and Rast, H. E. (1968). *Phys. Rev.* **168**, 964.
Mitra, S. S. (1963). *Phys. Rev.* **132**, 986.
Mitra, S. S. and Gielisse, P. J. (1964). Infrared spectra of crystals. *Progr. infrared Spectrosc.* **2**, 47.
Robertson, R., Fox, J. J., and Martin, A. E. (1934). *Phil. Trans. R. Soc.* **A232**, 463.
Shimanouchi, T., Tsuboi, M., and Miyazawa, T. (1961). *J. Chem. Phys.* **35**, 1597.
Siebert, H. (1954). *Z. anorg. allg. Chem.* **275**, 225.
Szigeti, B. (1950). *Proc. R. Soc.* **A204**, 51.
Szigeti, B. (1963). *J. Phys. Chem. Solids* **24**, 225.
Turner, W. J. and Reese, W. E. (1962). *Phys. Rev.* **127**, 126.
Wallis, R. F. and Maradudin, A. A. (1960). *Prog. theoret. Phys.* **34**, 1055.
Woods, A. D. B., Cochran, W., and Brockhouse, B. N. (1960). *Phys. Rev.* **119**, 980.

Author Index

Numbers in italics indicate the page on which the reference is listed

T

U

V

W

Y

Z

Subject Index

D

E

F

G

H

I

V

Y

Z